Charles Pendlebury

Lenses and Systems of Lenses

Charles Pendlebury

Lenses and Systems of Lenses

ISBN/EAN: 9783337034870

Printed in Europe, USA, Canada, Australia, Japan

Cover: Foto ©berggeist007 / pixelio.de

More available books at **www.hansebooks.com**

LENSES

AND

SYSTEMS OF LENSES,

TREATED AFTER THE MANNER OF GAUSS.

BY

CHARLES PENDLEBURY, M.A., F.R.A.S.,

SENIOR MATHEMATICAL MASTER OF ST. PAUL'S;
LATE SCHOLAR OF ST. JOHN'S COLLEGE, CAMBRIDGE.

CAMBRIDGE:
DEIGHTON, BELL AND CO.
LONDON: GEORGE BELL AND SONS.
1884.

In the following pages I have limited myself strictly to the consideration of that which is involved in the term "Lenses and Systems of Lenses"; and I have treated the matter in accordance with the methods of Gauss. But I hope that in course of time I may be able to extend the book so as to cover a wider area in the field of Geometrical Optics.

C.ʻ P.

St. Paul's School,
London.
February, 1884.

CONTENTS.

CHAPTER   PAGE

I.   REFRACTION AT A SINGLE SURFACE   .   .   .   1

II.   REFRACTION AT TWO SURFACES IN SUCCESSION   .   . 28

III.   REFRACTION AT ANY NUMBER OF SURFACES   .   . 52

IV.   ACHROMATISM   78

V.   THE DETERMINATION OF THE FOCI AND OF THE PRINCIPAL POINTS OF A SYSTEM OF LENSES. THE NODAL POINTS   80

VI.   THE DIFFERENT FORMS OF LENSES   86

Appendix. ON CONTINUED FRACTIONS   91

LENSES AND SYSTEMS OF LENSES.

CHAPTER I.

REFRACTION AT A SINGLE SURFACE.

1. IF a ray of light, or a pencil of rays traverse a system of coaxial lenses,—the lenses being of any thickness, of any focal lengths, and of any refractive indices whatever—the relation between the positions of the focus of the incident and the focus of the emergent pencil, and a formula for the *magnification* produced by the system of lenses, could formerly be determined only by an exceedingly cumbrous calculation. It was necessary, moreover, to repeat the process for each different system.

For the sake of simplicity it was often assumed that the lenses were indefinitely thin. The laboriousness of the calculations was thereby considerably reduced; but it is clearly a supposition which it is quite improper to make, except under very special circumstances.

In a paper communicated by Gauss to the Royal Society of Göttingen on the 10th of December, 1840,* it was shown how the solution of the problem could be made to depend upon the determination, for each system and once for all, of four fixed points situated upon the axis of the system. These points having been determined, the complete solution of the problem became a matter of simple algebra or Geometry.

---

* C. F. Gauss *Werke.* Band. V. Göttingen, 1840.

B

It is true that the calculation of the position of the four points is somewhat laborious, but the formulæ obtained are symmetrical, although long, and the formulæ for a system of $n+1$ lenses can be deduced very easily from that for a system of n lenses.

If therefore a table of formulæ be calculated for 2, 3, 4 ... lenses, which can easily be done, the application to any particular system is a question of Arithmetic and Algebra only.

Gauss' method is applicable to any system of coaxial lenses, whatever be the thicknesses of the lenses, whatever be the refractive indices of the media which occupy the spaces between them, and whether the medium in front of the first lens is the same as that behind the last, or not. The problem however becomes much simpler when these first and last media are the same.

One restriction, however, must be made. It is supposed that the angle which any ray makes with the axis, and also the distance from the axis of the point at which it cuts any refracting surface, are so small that their squares may be neglected. This is equivalent to neglecting *aberration*.

2. Let us consider a number of spherical surfaces, all of which have their centres of curvature upon a certain straight line. This straight line we may call the *axis* of the system.

If a system of this kind be intersected by a plane which contains the axis, the section will be such as is represented in fig. 1.

FIG. I.

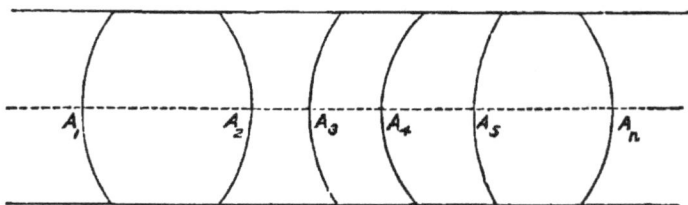

In this figure the straight line $A_1 A_2 ... A_n$, upon which all the centres of curvature are situated, is the axis of the system; and the points A_1, A_2, ...A_n in which the axis meets the successive surfaces may be called the *vertices* of the surfaces.

The surfaces may be of any number, of any degree of curvature, and at any distance apart; moreover, they may have their convex surfaces turned either way.

We will now suppose the spaces between each two consecutive surfaces to be occupied by homogeneous, not doubly refracting, media, of known refractive indices. The media may be all different, or two or more may be alike. But two similar media should not be adjacent to one another, for the effect would be the same as if the dividing surface were not there.

3. If we consider a particular case, and suppose the number of surfaces to be four, and the successive media to be air, glass, air, glass, and air, we so have the telescope in its simplest form, with one eyepiece and one object glass.

4. If now a ray of light proceed from a luminous point L, and cross the successive surfaces at the points P_1, P_2, P_3, ... P_n, its course will be bent at each of these points, but in consequence of our hypothesis concerning the nature of the media, its course between any two of the points will be a straight line. The path of the ray may therefore be represented by the broken line $LP_1P_2P_3 ... P_nL'$ (fig. 2).

FIG. 2.

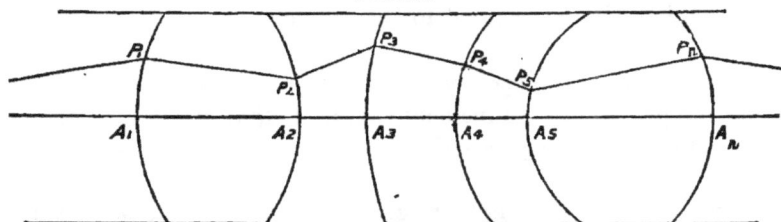

The end we aim at is the determination of a relation between the lines LP_1 and P_nL', so that when we know the initial path of a ray entering the system, we may at once be able to ascertain its final path on leaving it.

5. The general system includes also the case in which one or more of the surfaces are reflective. In order to make our results applicable to it, we have simply to consider the particular surface which is reflective, as if it were the boundary between two media whose refractive indices are μ_r and $-\mu_r$ respectively; μ_r being the refractive index of the medium immediately preceding the surface which reflects.

6. With regard to the rays themselves, we shall consider those only which are inclined at very small angles to the axis of the system, and which cross the surfaces at points very near the vertices.

If P be the point of incidence of a ray, A the vertex, and C the centre of curvature of the surface, it will follow that if the arc PA be small, so also will be the angle PCA (fig. 3).

FIG. 3.

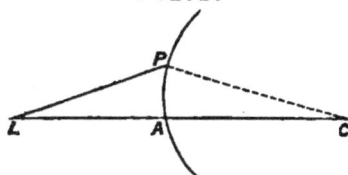

We shall suppose that the angle PCA is so small that all powers of its circular measure, higher than the second, may be neglected. Consequently, to this degree of approximation, we may consider the sine, tangent and circular measure of the angle PCA to be equal to one another.

7. In the first place, we will investigate formulæ connecting the position of the incident ray, with that of the ray after refraction, at *one surface only*.

8. *To find the angle between the incident ray and the refracted ray, after crossing one surface.*

Let C be the centre of curvature of any refracting surface, and A the vertex. Suppose XPX' to be the course of a ray which crosses the surface at the point P; and let the incident

ray and the refracted ray, produced if necessary, meet the
axis at the points X, X' respectively (fig. 4).

FIG 4.

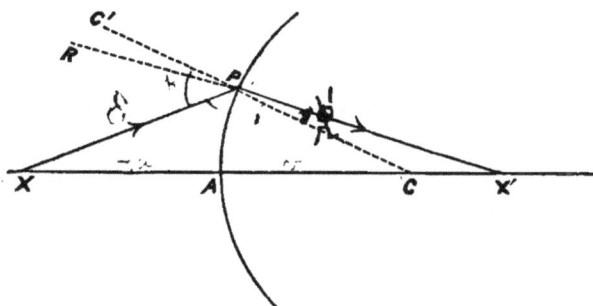

In this, and in all subsequent investigations, we will consider distances measured from left to right to be *positive*,
and all distances measured from right to left to be *negative*.
The above figure is so drawn that AC and AX', measured
from A are positive, and AX negative.

At the point P, the ray is bent through the angle XPR,
the point R being in $X'P$ produced. We will call this angle
the *Deviation* at P, and will denote it by δ.

Let the normal CP be produced to C', and let

$$\angle C'PX = \phi$$
and
$$\angle CPX' = \phi'$$,

then
$$\delta = \angle XPR$$

$$= \angle C'PX - \angle CPX'$$

$$= \phi - \phi'.$$

But if μ_0 and μ_1 be the refractive indices of the two
media, we have

$$\mu_0 \sin \phi = \mu_1 \sin \phi';$$

therefore by hypothesis

$$\mu_0 \phi = \mu_1 \phi';$$

therefore

$$\phi' = \frac{\mu_0 \phi}{\mu_1};$$

therefore
$$\delta = \phi - \frac{\mu_0 \phi}{\mu_1}$$

$$= \frac{\mu_1 - \mu_0}{\mu_1} \cdot \phi.$$

9. We may put the expression for the deviation in another form.

Let the arc $AP = h$, and denote the distances from A of X, X', and C, by $-u$, v, and r respectively; we then get

$$\delta = \frac{\mu_1 - \mu_0}{\mu_1} \phi$$

$$= \frac{\mu_1 - \mu_0}{\mu_1} \angle XPC'$$

$$= \frac{\mu_1 - \mu_0}{\mu_1} (\angle PXC + \angle PCX)$$

$$= \frac{\mu_1 - \mu_0}{\mu_1} \left(\frac{h}{r} - \frac{h}{u} \right) + Mh^3$$

$$= \frac{\mu_1 - \mu_0}{\mu_1} \left(\frac{h}{r} - \frac{h}{u} \right),$$

if we neglect h^3.

10. *If rays of light proceed from a luminous point X on the axis, they will, after crossing a refracting surface, meet again in one and the same point X', which also lies upon the axis. To find the relation between the positions of X and X'.*

With the same notation as before, we have (fig. 4)
$$\mu_0 \sin C'PX = \mu_1 \sin CPX';$$

therefore $\mu_0 \sin (PXA + PCA) = \mu_1 \sin (PCA - PX'A);$

therefore
$$\mu_0 \left(\frac{h}{r} - \frac{h}{u} \right) = \mu_1 \left(\frac{h}{r} - \frac{h}{v} \right) + Mh^3;$$

therefore
$$\frac{\mu_1}{v} - \frac{\mu_0}{u} = \frac{\mu_1 - \mu_0}{r} \quad \dots\dots\dots\dots\dots (1),$$

if we neglect h^3.

In this formula μ_0, μ_1, and r are quantities which depend only upon the nature of the media, and upon the curvature of the separating surface; they are the same whatever be the *ray* which we may consider. Hence we see, that since the formula gives only one value of v corresponding to a particular value of u, it follows that *all* the rays, which proceed from any point X on the axis, will after refraction meet again in a point X', also on the axis; the relative positions of these two points being given by equation (1).

Conversely, for a particular value of v we get from (1) only *one* value for u. Hence the rays, which after crossing a refracting surface meet together in a point X' on the axis, must before incidence have proceeded from one and the same point X, also lying upon the axis.

Again, if we consider X' instead of X as the luminous point, or the origin of the rays, and that the rays travel from right to left, it is clear that after refraction at the surface they will all meet together in the point X.

Consequently the point X bears the same relation to X', when X' is the origin of rays, that X' bears to X, when the origin of rays is at X.

Two points such as X and X', which are related to one another in this way, are called *Conjugate Points*.

11. There are now two particular cases to be considered :

(1) If the point X is at an infinite distance from A, we have $u = \infty$, and the incident rays are all parallel to the axis.

The equation connecting u and v being

$$\frac{\mu_1}{v} - \frac{\mu_0}{u} = \frac{\mu_1 - \mu_0}{r} \quad \ldots\ldots\ldots\ldots\ldots\ldots (1),$$

if we put $u = \infty$, we get

$$v = \frac{\mu_1 r}{\mu_1 - \mu_0} \quad \ldots\ldots\ldots\ldots\ldots\ldots\ldots (2).$$

(2) If the point X' be at an infinite distance from A, we have $v = \infty$; and the rays after refraction are parallel to the axis.

The corresponding value of u is given by the equation

$$u = \frac{\mu_0 r}{\mu_0 - \mu_1} \quad \ldots\ldots\ldots\ldots\ldots\ldots\ldots (3).$$

The two points determined by (2) and (3) are called the *Focal Points* or *Foci* of the given surface with respect to the given media, and are such that all rays whose paths in the first medium are parallel to the axis, will after refraction pass through the point given by (2); and all rays whose paths after refraction are parallel to the axis, must in the first medium have travelled in directions which, produced if necessary, would have passed through the point given by (3).

The Foci are commonly denoted by the letters F' and F. Their distances from the vertex A are called the *Focal Distances* of the surface, and are denoted by the symbols f' and f.

Hence
$$f = \frac{\mu_0 r}{\mu_0 - \mu_1},$$

$$f' = \frac{\mu_1 r}{\mu_1 - \mu_0},$$

and therefore
$$\mu_0 f' + \mu_1 f = 0.$$

If we introduce these symbols into equation (1), it becomes

$$\frac{f}{u} + \frac{f'}{v} = 1 \quad \dots\dots\dots\dots\dots\dots (4).$$

12. The equation $\quad \dfrac{f}{u} + \dfrac{f'}{v} = 1$

connects the distances of two conjugate points from the *vertex* of the refracting surface. It is sometimes convenient to have a relation between their distances from the *centre of curvature* of the surface, and sometimes between their distances from the *foci*. These relations we will now determine.

13. *To find a relation between the distances of two conjugate points from the centre of curvature of the refracting surface.*

Let C be the centre of curvature and A the vertex of the surface, X the origin of light, XP the path of a ray which meets the surface at P, and after refraction passes in the direction $X'P$. The refracted ray produced backwards meets the axis at the point X'. CPR is the normal at P (fig. 5).

FIG. 5.

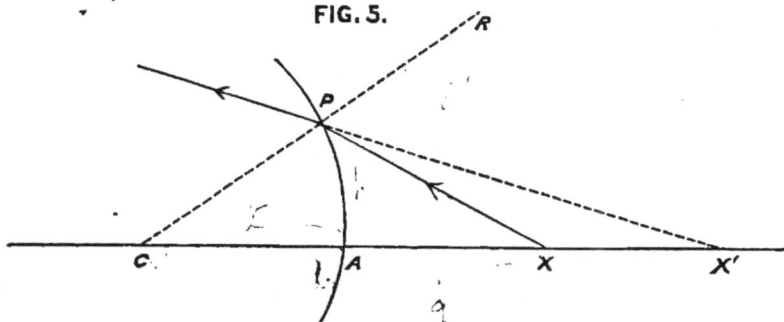

Let
$$\left. \begin{aligned} CX &= p \\ CX' &= q \\ CA &= r \\ \text{arc } AP &= h \end{aligned} \right\} \quad \text{and} \quad \left. \begin{aligned} \angle XPR &= \phi \\ \angle X'PR &= \phi' \end{aligned} \right\}.$$

Then we have $\qquad \mu_0 \sin XPR = \mu_1 \sin X'PR$;

therefore

$$\mu_0 \sin(PCA + PXA) = \mu_1 \sin(PCA + PX'A);$$

therefore $\qquad \mu_0 \left(\dfrac{h}{r} + \dfrac{h}{p-r} \right) = \mu_1 \left(\dfrac{h}{r} + \dfrac{h}{q-r} \right) + Mh^3.$

Hence, to the required degree of approximation, we have

$$\frac{\mu_0 p h}{r(p-r)} = \frac{\mu_1 q h}{r(q-r)};$$

therefore $\qquad p\mu_0 (q - r) = q\mu_1 (p - r);$

therefore $\qquad \dfrac{\mu_0}{q} - \dfrac{\mu_1}{p} = \dfrac{\mu_0 - \mu_1}{r} \quad \dotfill \quad (5),$

or $\qquad \dfrac{f}{p} + \dfrac{f}{q} = 1 \quad \dotfill \quad (6).$

The figure to which this article applies has been so drawn, that the distances from C of the various points might be all *positive*. Other cases may easily be deduced from this one.

14. *To find a relation between the distances of two conjugate points from the foci.*

$$
\begin{array}{lr}
\text{Let} & XF \; = d \\
\text{and} & X'F' = d'
\end{array} \Big\} \; .
$$

Then we have
$$d = u - f,$$
$$d' = v - f';$$

therefore
$$dd' = (u - f)(v - f')$$
$$= uv - uf' - vf + ff'$$
$$= ff' \quad \dotfill (7)$$

This relation, $dd' = ff'$, is called Newton's formula.

15. *To our degree of approximation, the tangent plane at the vertex may be considered as coincident with the surface itself.*

Let the incident and refracted rays meet the tangent plane at the points Q, Q' respectively; and let

$$
\begin{array}{l}
\angle PXA \; = \alpha \\
\angle PX'A = \alpha' \\
\angle PCA \; = \theta
\end{array} \Big\} \; \text{(fig. 6)}.
$$

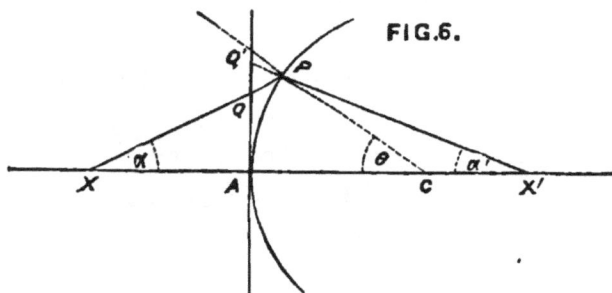

FIG.6.

Then we have
$$AQ \; = AX \tan \alpha,$$
$$AQ' = AX' \tan \alpha',$$

therefore
$$\frac{AQ}{AQ'} = \frac{AX \tan \alpha}{AX' \tan \alpha'}$$
$$= \frac{AX \sin \alpha}{AX' \sin \alpha'} \; \text{approximately}$$
$$= \frac{PX'}{AX'} \cdot \frac{AX}{PX} = 1,$$

since PX', PX differ from AX', AX only by quantities of the second order; therefore

$$AQ = AQ';$$

therefore the points Q, Q' and P coincide to our degree of approximation.

Hence the course of the ray may be represented by the broken line XQX' instead of by XPX'.

16. *If, through the foci F and F'', straight lines be drawn perpendicular to the axis, to meet the incident ray and the refracted ray in the points D and D' respectively, then*

$$FD + F'D' = AY,$$

Y being the point where the ray crosses the tangent plane at the vertex.[*]

Let
$$\left. \begin{array}{l} FD \ \ = z \\ F'D' = z' \\ AY \ \ = h \end{array} \right\} \text{(fig. 7)}.$$

FIG. 7.

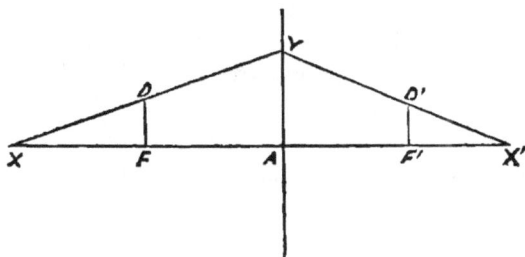

Then we have from (4)

$$\frac{f}{u} + \frac{f'}{v} = 1,$$

or

$$\frac{u-f}{u} + \frac{v-f'}{v} = 1;$$

therefore, by similar triangles, $\dfrac{z}{h} + \dfrac{z'}{h} = 1$,

or
$$z + z' = h.$$

---

* *Die Haupt- und Brenn- Puncte eines Linsen-systems*, von Carl Neumann. Leipzig. 1866.

17. If the ray XYX' goes through the vertex A, we have $h = 0$; therefore

$$z + z' = 0,$$

or $$z = -z'.$$

18. From the formula $z + z' = h$ we get a simple geometrical construction, whereby the path of the refracted ray may be determined when that of the incident ray is given; or, conversely.

If the incident ray be given, so are the distances FD and AY. Hence we can find $AY - FD$, which is equal to $F'D'$.

If, therefore, through the focus F' we draw the line $F'D'$ perpendicular to the axis and equal in length to $AY - FD$, we shall so determine the point D', which is a point on the refracted ray. The straight line YD' is then the path of the ray after refraction.

19. DEFINITION. If any number of rays proceed from the same luminous point, and be refracted in crossing a spherical surface which divides the first medium from another of different refractive index, the rays will after refraction meet again in one and the same point. Two such points, namely the point whence the rays proceed, and the point at which they meet again after refraction, are said to be *conjugate* to one another with respect to the media and the surface considered. They are called briefly *conjugate points*.

20. This definition involves a general theorem, of which we have not yet proved more than a particular case; that case, namely, in which the luminous point lies upon the axis.

We will next prove the theorem to hold for a luminous point which does not lie in the axis, but only for rays from it which lie in a plane passing through the axis.

Finally, we will consider the general case, and prove the theorem to be true for any position of the luminous point, and for rays from it in any directions whatever.

21. *Wherever the luminous point may be, all rays pro-ceeding from it in the plane which contains the luminous point and the axis, will after refraction meet together again in one and the same point.*

We will suppose the plane of the paper to contain the luminous point and the axis.

Let Q, chosen arbitrarily, be the position of a luminous point, and let QP_1Q', QP_2Q' be the paths of any two rays which proceed from it in the plane of the paper, and which meet the tangent plane at the vertex at P_1 and P_2 respectively; Q' being the point where they meet together after refraction.

Also let QP_3Q'' be a third ray which crosses the tangent plane at P_3 and meets the ray QP_1Q' at the point Q''.

We will show that the points Q' and Q'' coincide.

From Q, Q' and Q'' draw QN, $Q'N'$ and $Q''N''$ perpendicular to the axis, and meeting it at the points N, N', N'' respectively (fig. 8).

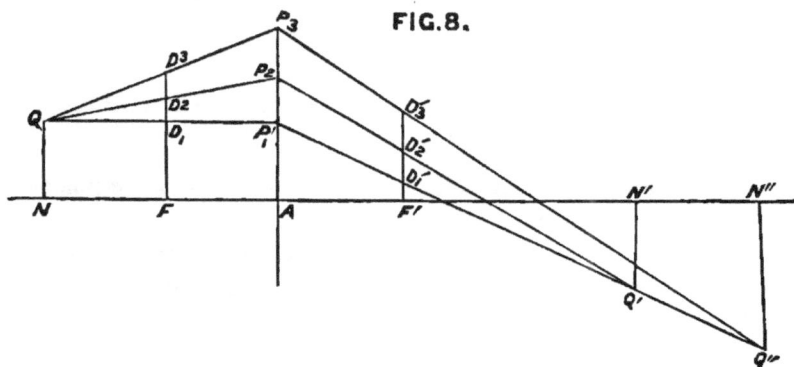

FIG. 8.

Through F and F' draw straight lines perpendicular to the axis, and meeting the rays at the points D_1, D_2, D_3 and D_1', D_2', D_3' respectively.

We have then

$$FD_1 + F'D_1' = AP_1,$$

and

$$FD_2 + F'D_2' = AP_2;$$

therefore, by subtraction,

$$D_1D_2 + D'_1D'_2 = P_1P_2;$$

therefore

$$\frac{D_1D_2}{P_1P_2} + \frac{D'_1D'_2}{P_1P_2} = 1;$$

therefore, by similar triangles,

$$\frac{QD_1}{QP_1} + \frac{Q'D'_1}{Q'P_1} = 1;$$

therefore

$$\frac{NF}{NA} + \frac{N'F'}{N'A} = 1;$$

therefore

$$\frac{NA - NF}{NA} + \frac{N'A - N'F'}{N'A} = 1;$$

therefore

$$\frac{FA}{NA} + \frac{F'A}{N'A} = 1;$$

therefore

$$\frac{f}{NA} + \frac{f'}{N'A} = 1.$$

In an exactly similar way, by considering the two rays QP_1Q'' and QP_2Q'', we shall get

$$\frac{f}{NA} + \frac{f'}{N''A} = 1.$$

Comparing these two results, we see at once that

$$N'A = N''A.$$

Hence the points N' and N'', and consequently the points Q' and Q'', coincide.

Wherefore all rays which proceed from any luminous point Q, will after refraction meet again in one and the same point Q'.*

22. From the result of the preceding article we obtain a simple geometrical construction, whereby we may determine the position of a point conjugate to a given one.

Let Q be the given point. Then we know that the point conjugate to Q is the point of concurrence of *all* the rays

---

* *Die Haupt- und Brenn- Puncte eines Linsen Systems*, von Carl Neumann. Leipzig. 1866.

which proceed from Q. Hence it will be enough to determine the point of concurrence of any two.

It happens that there are two rays, whose paths we know from beginning to end ; namely, the ray which proceeds from Q in a direction parallel to the axis, and the ray, which passes through F.

The former of these passes through F' after refraction, and the latter after refraction travels in a direction parallel to the axis.

Hence we have the following geometrical construction for the determination of the point conjugate to Q (fig. 9).

FIG. 9.

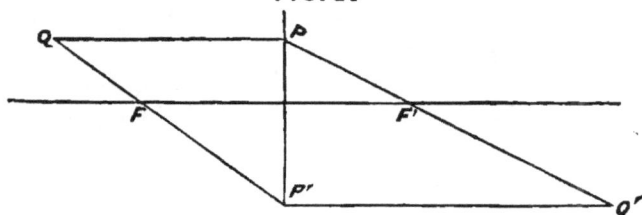

From Q draw QP parallel to the axis and meeting the tangent plane at the vertex at P. Join PF', and produce it. Again, join QF and produce it to meet the tangent plane at P'. Through P' draw $P'Q'$ parallel to the axis, and let it meet PF' produced at Q'.

Then, since the rays QP and QF meet after refraction at the point Q', it follows that Q' is the point conjugate to Q.

23. In Arts. 10 and 21 have been proved two particular cases of our fundamental theorem. We will now consider the general case, in which the position of the luminous point, and the directions of the rays, are both unrestricted. For this purpose it will be necessary to use the algebraic equations to the straight line in Three Dimensions.*

24. We will take the axis of the surface for the axis of x, and any plane perpendicular to it for the plane of yz.

---

* *Verdet Œuvres*, tome IV. part ii. Conférences de Physique.

The equations of the incident ray may then be written in
the form

$$y = mx + b \atop z = nx + c \Big\},$$

or, more conveniently, in the equally general form

$$y = \frac{m_0}{\mu_0}(x - a) + b_0 \atop z = \frac{n_0}{\mu_0}(x - a) + c_0 \Bigg\},$$

where μ_0 is the index of refraction of the first medium;

$\frac{m_0}{\mu_0}$, $\frac{n_0}{\mu_0}$ the tangents of the angles of inclination to the axis
of x of the projections of the incident ray on the
planes of xy and xz respectively;

a the abscissa of the vertex of the refracting surface;

a, b_0, c_0 the coordinates of the point at which the ray meets
the tangent plane at the vertex.

In a similar manner the refracted ray may be represented
by the equations

$$y = \frac{m_1}{\mu_1}(x - a) + b_1 \atop z = \frac{n_1}{\mu_1}(x - a) + c_1 \Bigg\},$$

μ_1 being the index of refraction of the second medium.

It follows from our hypothesis, that $\frac{m_0}{\mu}$, $\frac{n_0}{\mu_0}$, $\frac{m_1}{\mu_1}$, $\frac{n_1}{\mu_1}$, b_0,
c_0, b_1 and c_1 are all small quantities of the first order.

25. We will first investigate the relation between b_1 and
b_0, and between c_1 and c_0.

Let P be the point of incidence,

C the centre and r the radius of curvature of the
refracting surface,

A the vertex;

and let a plane drawn through C parallel to the plane zy
be met by the incident and refracted rays at T and T'
respectively. From the fact that the incident ray, the

refracted ray, and PC the normal at P, are all in the same plane, it follows that $CT'T$ is a straight line (fig. 10).

FIG.10.

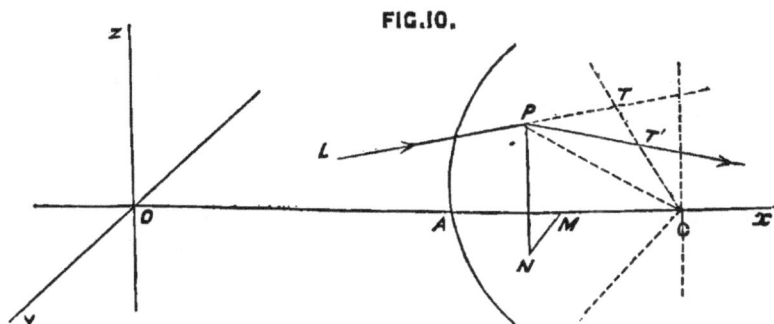

Let
$$\angle PTC = \alpha,$$
$$\angle PT'C = \alpha',$$
$$\angle TPC = \phi_0,$$
$$\angle T'PC = \phi_1,$$
$$\angle PCA = \theta,$$
$$CP = r,$$
$$OA = a,$$

and let x, y, and z be the coordinates of P.

Draw PN, NM perpendicular to the planes xy, xz respectively.

Then
$$OM = OA + AM$$
$$= a + r (1 - \cos\theta).$$

Now, at the point P which is common to the incident and refracted rays, the expressions for y given by their respective equations must be the same, hence

$$\frac{m_0}{\mu_0}(x - a) + b_0 = \frac{m_1}{\mu_1}(x - a) + b_1,$$

where
$$x = OM.$$

Hence, substituting for x the value found for it above, we get

$$\frac{m_0 r}{\mu_0}(1 - \cos\theta) + b_0 = \frac{m_1 r}{\mu_1}(1 - \cos\theta) + b_1;$$

C

therefore $\quad b_1 = b_0 + r(1 - \cos\theta)\left[\dfrac{m_0}{\mu_0} - \dfrac{m_1}{\mu_1}\right].$

In accordance with our hypotheses $\dfrac{m_0}{\mu_0}$ and $\dfrac{m_1}{\mu_1}$ are small quantities of the *first* order, and $1 - \cos\theta$ is a small quantity of the *second* order, consequently

$$r(1 - \cos\theta)\left(\frac{m_0}{\mu_0} - \frac{m_1}{\mu_1}\right)$$

is a small quantity of the *third* order and may be neglected.

Therefore, to this degree of approximation, we get

$$b_1 = b_0;$$

and in a similar way it may be shown that

$$c_1 = c_0;$$

therefore the equations to the refracted ray may be written in the form

$$\left.\begin{aligned} y &= \frac{m_1}{\mu_1}(x - a) + b_0 \\[2mm] z &= \frac{n_1}{\mu_1}(x - a) + c_0 \end{aligned}\right\}.$$

26. Next, we will determine the relations between m_1, n_1, and m_0, n_0.

We have from the figure

$$\frac{CT'}{CT} = \frac{CT'}{CP} \cdot \frac{CP}{CT}$$

$$= \frac{\sin\phi_1}{\sin\alpha'} \cdot \frac{\sin\alpha}{\sin\phi_0}$$

$$= \frac{\mu_0 \sin\alpha}{\mu_1 \sin\alpha'}.$$

But the ratio $\dfrac{CT'}{CT}$ is also equal to the ratio of the y coordinates of the points T' and T, *i.e.* of the points on the two portions of the ray whose abscissa is $a + r$; therefore

$$\frac{\mu_0 \sin\alpha}{\mu_1 \sin\alpha'} = \frac{-\dfrac{m_1 r}{\mu_1} + b_0}{\dfrac{m_0 r}{\mu_0} + b_0};$$

therefore $\qquad \dfrac{m_1 r}{\mu_1} + b_0 = \dfrac{\mu_0 \sin \alpha}{\mu_1 \sin \alpha'} \left(\dfrac{m_0 r}{\mu_0} + b_0 \right).$

But α and α' are both very nearly right angles; therefore $\dfrac{\sin \alpha}{\sin \alpha'}$, differs from unity by a small quantity of the *second* order. Also $\dfrac{m_0 r}{\mu_0} + b_0$ is a small quantity of the *first* order; therefore, if we neglect small quantities of the *third* order, we get

$$\frac{m_1 r}{\mu_1} + b_0 = \frac{\mu_0}{\mu_1} \left(\frac{m_0 r}{\mu_0} + b_0 \right);$$

therefore $\qquad \dfrac{m_1 r}{\mu_1} = \dfrac{m_0 r}{\mu_1} + b_0 \left(\dfrac{\mu_0}{\mu_1} - 1 \right);$

therefore $\qquad m_1 = m_0 + \dfrac{\mu_0 - \mu_1}{r} b_0.$

In a similar way it may be shown that

$$n_1 = n_0 + \frac{\mu_0 - \mu_1}{r} c_0,$$

subject to the condition, that we may neglect small quantities of the *third* order.

27. We have now found expressions for the constants involved in the equations to the refracted ray, in terms of the constants involved in the equations to the incident ray. It follows that if the equations to the incident ray be

$$\left. \begin{array}{l} y = \dfrac{m_0}{\mu_0} (x - a) + b_0 \\[2mm] z = \dfrac{n_0}{\mu_0} (x - a) + c_0 \end{array} \right\},$$

then the equations to the refracted ray will be

$$\left. \begin{array}{l} y = \left(m_0 + b_0 \dfrac{\mu_0 - \mu_1}{r} \right) \dfrac{1}{\mu_1} (x - a) + b_0 \\[3mm] z = \left(n_0 + c_0 \dfrac{\mu_0 - \mu_1}{r} \right) \dfrac{1}{\mu_1} (x - a) + c_0 \end{array} \right\}.$$

28. With the help of these equations may be proved the most general case of the proposition, that

All rays which proceed from one and the same point will, after refraction, meet again in one and the same point.

Let ξ, η, ζ be any point on the incident ray, and ξ', η', ζ' a point on the refracted ray; then we have

$$\left. \begin{array}{l} \eta = \dfrac{m_0}{\mu_0}(\xi - a) + b_0 \\[3mm] \eta' = \left(m_0 + b_0 \dfrac{\mu_0 - \mu_1}{r}\right) \dfrac{1}{\mu_1}(\xi' - a) + b_0 \end{array} \right\}.$$

Eliminating b_0 from these two equations, we get

$$\eta' - \frac{m_0}{\mu_1}(\xi' - a) = \left[\eta - \frac{m_0}{\mu_0}(\xi - a)\right]\left[\frac{\mu_0 - \mu_1}{\mu_1 r}(\xi' - a) + 1\right];$$

therefore

$$\eta' - \eta\left[\frac{\mu_0 - \mu_1}{\mu_1 r}(\xi' - a) + 1\right]$$
$$= m_0\left[\frac{\xi' - a}{\mu_1} - \frac{\xi - a}{\mu_0}\left\{\frac{\mu_0 - \mu_1}{\mu_1 r}(\xi' - a) + 1\right\}\right].$$

In a similar manner it may be shown, that

$$\zeta' - \zeta\left[\frac{\mu_0 - \mu_1}{\mu_1 r}(\xi' - a) + 1\right]$$
$$= n_0\left[\frac{\xi' - a}{\mu_1} - \frac{\xi - a}{\mu_0}\left\{\frac{\mu_0 - \mu_1}{\mu_1 r}(\xi' - a) + 1\right\}\right].$$

Each of these relations between ξ', η', ζ' and ξ, η, ζ involves an unknown parameter in the first degree, m_0 in the first relation and n_0 in the second. These quantities depend upon the particular ray considered. Each is constant so long as we consider the same ray, but assumes a different value if we change to another ray.

Now it will be noticed that the coefficient of m_0 is the same as the coefficient of n_0. If we equate this coefficient to zero, we get

$$\frac{\xi' - a}{\mu_1} = \frac{\xi - a}{\mu_0}\left\{\frac{\mu_0 - \mu_1}{\mu_1 r}(\xi' - a) + 1\right\} \quad \ldots\ldots\ldots\ldots \text{(i)},$$

and consequently

$$\eta' = \eta \left\{ \frac{\mu_0 - \mu_1}{\mu_1 r} (\xi' - a) + 1 \right\} \quad \dots\dots\dots\dots \text{(ii)},$$

and

$$\zeta' = \zeta \left\{ \frac{\mu_0 - \mu_1}{\mu_1 r} (\xi' - a) + 1 \right\} \quad \dots\dots\dots\dots \text{(iii)}.$$

The constants which are involved in these three equations, namely, a, μ_0, μ_1, r, depend entirely upon the nature of the media, and on the position and curvature of the refracting surface. They are therefore *absolute* constants.

It follows that these equations hold for *any ray whatever* that before incidence passed through the point $\xi\eta\zeta$.

From (i) we get

$$\frac{\xi' - a}{\mu_1} \left\{ 1 - \frac{\mu_0 - \mu_1}{r} \frac{\xi - a}{\mu_0} \right\} = \frac{\xi - a}{\mu_0} \, ;$$

therefore

$$\xi' = a + \frac{\mu_1 r (\xi - a)}{\mu_0 r - (\mu_0 - \mu_1)(\xi - a)} \, ,$$

which gives us ξ' in terms of ξ.

If we substitute this value of ξ' in equations (ii) and (iii), we get η' in terms of ξ, η, and ζ' in terms of ξ, ζ.

Hence we see that we have found a point $\xi'\eta'\zeta'$ on the refracted ray, whose position depends solely upon that of $\xi\eta\zeta$, and is independent of the particular ray considered. Consequently, *every* ray which before incidence passes through the point whose coordinates are ξ, η, ζ, will after refraction pass through the point whose coordinates ξ', η', ζ' are determined by the equations (i), (ii) and (iii), whether the incident ray meets the axis or not.

This proves the theorem completely.

29. Equation (i) in the preceding article leads to a very important geometrical property of conjugate points. It shows us that ξ' depends upon ξ *alone*, and is independent of η and ζ. Hence, to all points which have the same ξ, correspond conjugate points, all of which have the same ξ'. Consequently, if any number of points lie upon a plane perpendicular to the axis, the points conjugate to them will also all lie upon a plane perpendicular to the axis.

Two such planes are said to be conjugate to one another, and are called briefly *Conjugate Planes*.

The property of conjugate planes may however be proved in an elementary manner in the case in which the plane of incidence contains the axis of the refracting surface. We will assume that the plane of incidence is coincident with the plane of the paper.

30. *To investigate formulæ for the Geometrical determination of the position of the point conjugate to a given one.*

Let P, P' be two points conjugate to one another. The position of P being given, it is required to find that of P' (fig. 11).

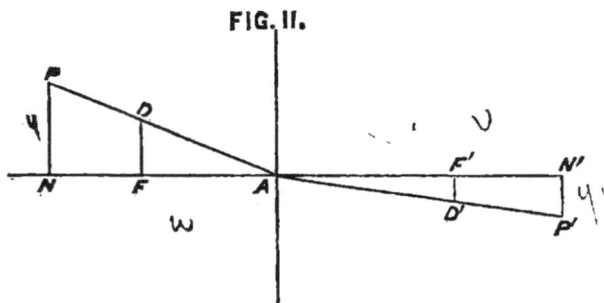

FIG. II.

We know that PAP' will be the course of the ray which passes through the vertex A.

Draw PN, $P'N'$ perpendicular to the axis and meeting it in N, N' respectively; and through F and F' let FD, $F'D'$ be drawn perpendicular to the axis to meet the incident and refracted rays respectively in D and D'.

From the formula $z + z' = h$ (Art. 16) we have, since $h = 0$,

$$z = -z';$$

therefore FD and $F'D'$ are equal in length, but on opposite sides of the axis; therefore, by similar triangles,

$$\frac{FD}{PN} = \frac{AF}{AN} = \frac{f}{AN},$$

and also

$$\frac{F'D'}{P'N'} = \frac{AF'}{AN'} = \frac{f'}{AN'}.$$

Hence $\quad \dfrac{f \cdot PN}{AN} = \dfrac{f' \cdot P'N'}{AN'}$, numerically,

and we have previously shown (Art. 21) that

$$\frac{f}{AN} + \frac{f'}{AN'} = 1.$$

These equations are sufficient to determine AN' and $P'N'$ when AN and PN are known.

If we put u, y, v, y' for AN, PN, AN', $P'N'$ respectively, the equations may be written in the form

$$\left.\begin{array}{c} \dfrac{fy}{u} = \dfrac{f'y'}{v} \\[2mm] \dfrac{f}{u} + \dfrac{f'}{v} = 1 \end{array}\right\} \quad \begin{array}{l} \dots\dots\dots\dots\dots\text{(i),} \\[2mm] \dots\dots\dots\dots\dots\text{(ii).} \end{array}$$

31. Equation (ii) of the preceding article leads at once to the property of conjugate planes, which, in Art. 29, we deduced from the formulæ in the general theorem. The equation shows that the value of v depends only upon that of u, and is independent of y.

Hence, if any number of points be taken, lying upon a plane perpendicular to the axis of the refracting surface, the points conjugate to them will also lie upon a plane perpendicular to the axis.

The distances u, v of two conjugate planes from the vertex are connected by the equation

$$\frac{f}{u} + \frac{f'}{v} = 1.$$

32. In Art. 22 it was shown how the conjugate point might be found graphically by following the paths of two known rays to their point of intersection. Now, however, by means of the equation

$$\frac{f}{u} + \frac{f'}{v} = 1,$$

we may determine the plane conjugate to that which passes through the luminous point; and for the determination of the *point* conjugate to the luminous point, we need follow only *one* ray to its intersection with the plane.

The ray most suitable for the purpose is that which passes through the centre of curvature of the refracting surface. This ray crosses the surface without undergoing refraction, and its path from beginning to end is in one straight line (fig. 12).

FIG. 12.

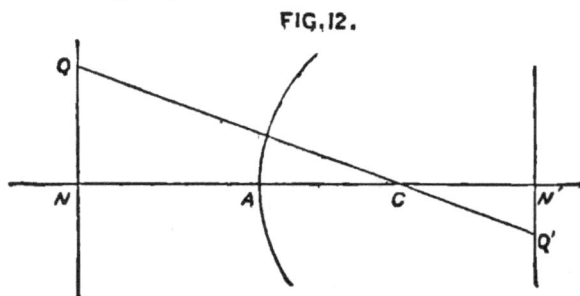

Let Q be the luminous point, C the centre of curvature, and QN the plane perpendicular to the axis which contains Q, and let $Q'N'$ be the plane conjugate to QN.

Join QC, and produce it to meet the plane $Q'N'$ at Q'.

Then Q' is the point which is conjugate to Q.

Hence we have the general theorem that *all straight lines which join points to their conjugates pass through the centre of curvature of the refracting surface.*

33. *Definition.* The point Q', through which pass after refraction all the rays which proceed from Q, is called the *image* of Q.

34. Let us suppose that there are a number of points such as Q, all of which lie on the plane QN. To each of these will correspond an *image* lying on the plane $Q'N'$. And if the points Q form in the aggregate an object of finite size and definite shape, the points Q' will form an image exactly similar in shape to the original object formed by the points Q.

The *size* of the image will not, however, be the same

as that of the object. It is necessary therefore to determine the relation between them.

35. *Definition*. The ratio of the linear dimensions of the image to the linear dimensions of the object is called the *magnification*, or the magnifying power of the surface.

36. *To find an expression for the magnification caused by refraction at a surface.*

Let Q be a point on the boundary of the object, then the conjugate point Q' will be a point on the boundary of the image (fig. 12).

Consequently the magnification is represented by the ratio of $Q'N'$ to QN.

This ratio we will denote by m, and we will find an expression for m in terms of u and v, the distances respectively of N and N' from the vertex A.

We have
$$m = \frac{Q'N'}{QN}$$
$$= \frac{CN'}{CN} \text{ by similar triangles,}$$
$$= \frac{v-r}{u-r}.$$

But
$$\frac{\mu_1}{v} - \frac{\mu_0}{u} = \frac{\mu_1 - \mu_0}{r};$$

therefore
$$\mu_1 u\,(r-v) = \mu_0 v\,(r-u);$$

therefore
$$\frac{v-r}{u-r} = \frac{\mu_0 v}{\mu_1 u};$$

therefore
$$m = \frac{\mu_0 v}{\mu_1 u} \quad\text{......................(1).}$$

37. We will now introduce a notation which will subsequently be found very useful.

The symbols u and v denote what may be called the *absolute* distances of N and N' from A.

Let us denote $\dfrac{u}{\mu_0}$ and $\dfrac{v}{\mu_1}$ by u' and v' respectively, then the formula for the magnification becomes

$$m = \frac{v'}{u'} \quad\dotfill(2).$$

We may call u' and v' the *reduced* distances of N and N' from A.

The same notation may be adopted for all linear magnitudes, any *reduced* distance being obtained by dividing the corresponding *absolute* distance by the refractive index of the medium in which it is measured.

38. *Helmholtz' formula for magnification.**

Let α and α' be the angles at which an incident ray XY and the refracted ray YX' are respectively inclined to the axis (fig. 13).

FIG. 13.

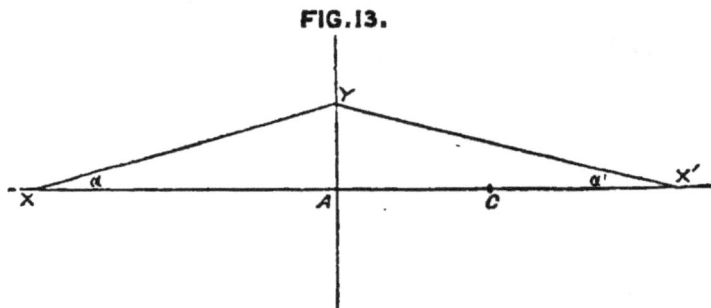

Then $u \tan \alpha = AY = v \tan \alpha'$;

therefore $m = \dfrac{\mu_0 v}{\mu_1 u}$

$$= \frac{\mu_0 \tan \alpha}{\mu_1 \tan \alpha'} \quad\dotfill(3).$$

This formula is an important one, for it connects the magnification with the angle of divergence of the rays, and is independent of the curvature of the refracting surface.

---

* Helmholtz, *Optique Physiologique.*

39. It may be noticed that an expression for the magnification may also be obtained from equations (ii) and (iii) of Article 28.

We get
$$m = \frac{\eta'}{\eta} = \frac{\zeta'}{\zeta}$$

$$= \frac{\mu_0 - \mu_1}{\mu_1 r} (\xi' - a) + 1,$$

where
$$\xi' = a + \frac{\mu_1 r (\xi - a)}{\mu_0 r - (\mu_0 - \mu_1) (\xi - a)}.$$

Hence

$$m = \frac{\mu_0 - \mu_1}{\mu_1 r} \frac{\mu_1 r (\xi - a)}{\mu_0 r - (\mu_0 - \mu_1) (\xi - a)} + 1$$

$$= \frac{(\mu_0 - \mu_1)(\xi - a)}{\mu_0 r - (\mu_0 - \mu_1)(\xi - a)} + 1$$

$$= \frac{\mu_0 r}{\mu_0 r - (\mu_0 - \mu_1)(\xi - a)} \quad \dots\dots\dots\dots(4).$$

CHAPTER II.

40. Many of the properties which belong to a ray of light when refracted at one surface only, may be extended almost directly to the case in which the ray is bent a second time in traversing a second surface.

We assume that the second surface is related in position to the first in the manner described in fig. 1; that it is wholly independent of it as far as curvature is concerned; and that the two surfaces are at any distance whatever apart.

Two such surfaces combined form an ordinary thick lens, the character of which depends upon the curvatures of the two surfaces, the directions in which their concavities are turned, and the refractive index of the medium between them.

The form of what we may call our *standard* lens is given in fig. 14; for, in accordance with our convention, both the radii of curvature are there positive. It is therefore the simplest case to demonstrate, as well as the one from which particular cases can most easily be deduced.

FIG. 14.

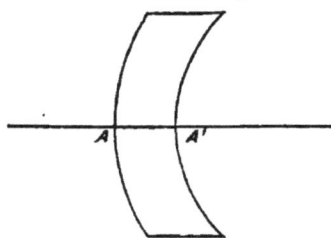

We will also assume, as a rule, that the first and last media are similar—air, for example—and we will denote the refractive indices of the successive media by μ_0, μ_1, μ_0 respectively.

If, then, r_1 and r_2 be the radii of curvature of the two surfaces, the properties of any particular lens may be deduced from this general case by assigning to r_1 and r_2 their proper values, and to μ_1 the value of the refractive index of the particular material.

41. *If rays proceed from a luminous point and traverse two refracting surfaces in succession, they will, after emergence, meet again in one and the same point.*

Suppose P to be the luminous point.

It has been proved in Chap. I. that the rays from P will, after refraction at the *first* surface, meet again in a certain point P_1.

The point P_1, or the image of P with respect to the first surface, may be considered as a source of light from which rays proceed across the *second* surface.

All these rays, after refraction at the second surface, will, in consequence of the same law, meet again in a certain point P', the point P' being the image of P_1 with respect to the second surface.

Hence, all rays which proceed from a luminous point P and traverse two refracting surfaces in succession will, on emergence, meet again in one and the same point P'. This result holds whatever be the position of P, and whether the plane of incidence contain the axis of the lens or not.

42. It is obvious that, just as in the case of one surface only, P and P' are reciprocally related to one another, and that if we were to consider P' as the source of light, all rays from it which traverse the two surfaces would on emergence meet together at the point P.

Hence P and P' are *conjugate* to one another with respect to the lens considered; or, in other words, P' is the image of P.

43. The definition of conjugate points leads directly to the two following propositions:

(i) If s and σ denote any two incident rays, and s' and σ' the corresponding emergent rays, the point of concurrence of s and σ is conjugate to the point of concurrence of s' and σ'.

(ii) If P and Q be a pair of conjugate points, and p and q another pair, a ray which before incidence passes through P and p will after emergence pass through Q and q.

44. *If a number of points P lie upon a plane perpendicular to the axis, all the points P' conjugate to them will also lie upon a plane perpendicular to the axis.*

For the points which are conjugate to the system P with respect to the first surface lie upon a certain plane perpendicular to the axis. This was proved in the former chapter.

We will call this system of points P_1, and we may consider the points as sources of light from which rays traverse the second surface.

Again, we know that all the points P' which are conjugate to the points P_1 with respect to the second surface also lie upon a certain plane perpendicular to the axis.

But the points P' are conjugate to the points P with respect to the lens.

Whence the proposition follows.

45. If P and P' be conjugate points, and planes pass through them perpendicular to the axis, it follows that any point on one of the planes has its conjugate on the other. Two such planes are said to be conjugate to one another with respect to the lens, and are called briefly *Conjugate Planes*.

Also the points where the planes meet the axis of the lens are called *Conjugate Foci*.

46. *If any two conjugate planes be taken, and any number of points on one plane be joined to their conjugate points on the other, all these straight lines will meet the axis in the same point.*

Let PN, P_1N_1, $P'N'$ be planes such that PN and P_1N_1 are conjugate to one another with respect to the first surface, and P_1N_1 and $P'N'$ conjugate with respect to the second surface; and let C_1, C_2 be the centres of curvature of the two surfaces respectively (fig. 15).

FIG . 15.

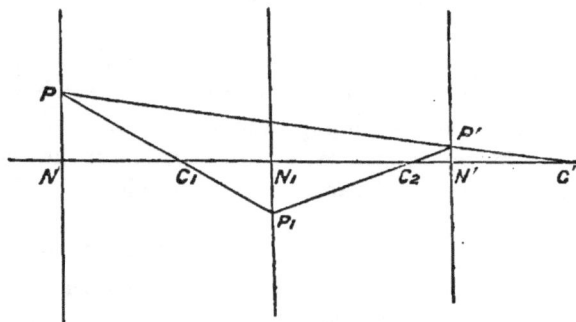

Let P' be conjugate to P with respect to the lens, and let the straight line PP' meet the axis at the point C'. We will show that C' is a fixed point, for different positions of P in the plane PN.

If P_1 be conjugate to P with respect to the first surface, and therefore conjugate to P' with respect to the second surface, it has been proved in Art. 32, that the straight lines PP_1 and P_1P' pass through C_1 and C_2 respectively.

We have
$$\frac{PN}{P_1N_1} = \frac{NC_1}{N_1C_1} = \text{constant},$$

and
$$\frac{P_1N_1}{P'N'} = \frac{N_1C_2}{N'C_2} = \text{constant};$$

therefore
$$\frac{PN}{P'N'} = \text{constant};$$

therefore
$$\frac{NC'}{N'C'} = \text{constant};$$

therefore
$$\frac{NN'}{N'C'} = \text{constant};$$

therefore
$$N'C' = \text{constant};$$

therefore C' is a fixed point for all positions of P in the plane PN.

47. If the point N move along the axis to an infinite distance from the lens, the rays which proceed from it, in the limiting position, will before incidence be parallel to the axis, and after emergence will meet at a certain point N' on the axis.

Again, if the point N' move along the axis to an infinite distance from the lens, the emergent rays which converge to N' will, in the limiting position, be parallel to the axis, and must before incidence have issued from an origin of light at a point N situated upon the axis.

The limiting position of N' as N moves off to an infinite distance, and the limiting position of N as N' moves off to an infinite distance, are called the *Principal Foci* of the lens. They are commonly referred to simply as *the* Foci, and are denoted by the letters F' and F respectively.

Hence all rays which before incidence are parallel to the axis will after refraction pass through the point F', and all rays which after emergence are parallel to the axis must have proceeded before incidence from the point F.

48. If we denote by I and I' respectively the infinitely distant points towards which N and N' move, it follows that I and F', and F and I' are pairs of conjugate points.

49. The planes through the foci F and F' perpendicular to the axis are called the *Focal Planes*.

50. The planes conjugate to the Focal Planes are at an infinite distance.

Hence, if the luminous point be on a Focal Plane, it follows that all the rays which proceed from it will, on emergence, be parallel to one another.

Also, if an image fall on a Focal Plane, it follows in the same way that the incident rays must all have been parallel to one another.

51. In the case of refraction at *one* surface only it was shown that the distances u and v of conjugate foci from the vertex of the surface are connected by the equation

$$\frac{\mu_1}{v} - \frac{\mu_0}{u} = \frac{\mu_1 - \mu_0}{r},$$

or by

$$\frac{f}{u} + \frac{f'}{v} = 1.$$

We will now investigate the corresponding formula for the case of two surfaces.

52. *To find the relation between the positions of conjugate foci when a ray is refracted through a lens.*

Let $QPP'X_3$ be the path of a ray, which crosses the surfaces at the points P, P' respectively, and let X_1, X_2 be the points at which the portions QP, PP', produced if necessary, meet the axis: X_3 being the point at which the axis is met by the emergent ray.

FIG. 16.

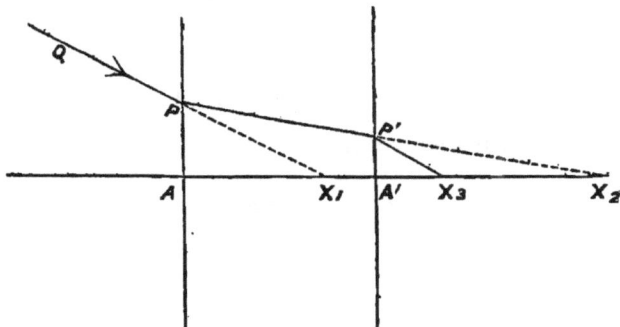

Then X_1, X_2 are conjugate to one another with respect to the first surface; and X_2, X_3 are conjugate with respect to the second surface.

Let A and A' be the two vertices. The thickness of the lens, AA', is a positive quantity, but we may represent it by $-t$, if we consider t to be itself negative. With this notation we shall make the formulæ more symmetrical.

D

Let u, v_1, be the distances of X_1, X_2 respectively from the vertex A;

$v_1 + t$, v, the distances of X_3, X_3 respectively from the vertex A';

r, s, the radii of the first and second surfaces respectively.

Then we have

$$\frac{\mu_1}{v_1} - \frac{\mu_0}{u} = \frac{\mu_1 - \mu_0}{r},$$

and also

$$\frac{\mu_0}{v} - \frac{\mu_1}{v_1 + t} = \frac{\mu_0 - \mu_1}{s}.$$

For simplicity we will denote $\dfrac{\mu_1 - \mu_0}{r}$ by ρ and $\dfrac{\mu_0 - \mu_1}{s}$ by σ. Then we get

$$\frac{\mu_1}{v_1} - \frac{\mu_0}{u} = \rho,$$

$$\frac{\mu_0}{v} - \frac{\mu_1}{v_1 + t} = \sigma.$$

These equations may be simplified still further by using *reduced* distances instead of *absolute* distances; as explained in Art. 37.

We will write u' for $\dfrac{u}{\mu_0}$, v_1' for $\dfrac{v_1}{\mu_1}$, t' for $\dfrac{t}{\mu_1}$, and so on.

The equations then become

$$\frac{1}{v_1'} - \frac{1}{u'} = \rho,$$

$$\frac{1}{v'} - \frac{1}{v_1' + t'} = \sigma.$$

It will however be unnecessary to use the accents if we remember that in future the symbols represent *reduced* and not *absolute* distances.

So we get

$$\frac{1}{v_1} - \frac{1}{u} = \rho,$$

$$\frac{1}{v} - \frac{1}{v_1 + t} = \sigma.$$

Hence
$$v = \cfrac{1}{\sigma + \cfrac{1}{v_{_1} + t}}$$

$$= \cfrac{1}{\sigma + \cfrac{1}{t + \cfrac{1}{\rho + \cfrac{1}{u}}}}$$

$$= \frac{1}{\sigma +} \frac{1}{t +} \frac{1}{\rho +} \frac{1}{u} \dots\dots\dots\dots\dots \text{ (i),}$$

or
$$v = \frac{u\,(\rho t + 1) + t}{u\,(\sigma \rho t + \sigma + \rho) + \sigma t + 1}$$

$$= \frac{Cu + D}{Au + B} \dots\dots\dots\dots\dots\dots\dots\dots\dots\text{(ii),}$$

where
$$A = \sigma \rho t + \rho + \sigma,$$
$$B = \sigma t + 1,$$
$$C = \rho t + 1,$$
$$D = t.$$

53. We may notice that the constants B, C and D can be expressed in terms of A.

For
$$B = \frac{dA}{d\rho},$$

$$C = \frac{dA}{d\sigma},$$

$$D = \frac{d^2 A}{d\rho\, d\sigma}.$$

54. The constants A, B, C, D are also connected by the equation

$$AD - BC = (\sigma \rho t + \rho + \sigma)\, t - (\sigma t + 1)\,(\rho t + 1)$$
$$= \sigma \rho t^2 + \rho t + \sigma t - (\sigma \rho t^2 + \sigma t + \rho t + 1)$$
$$= -1.$$

55. If in the formula

$$v = \frac{Cu + D}{Au + B}$$

we put $u = \infty$, we get

$$A'F' = v = \frac{C}{A} \quad\dots\dots\dots\dots\dots\text{(i)},$$

and if we put $v = \infty$, we get

$$AF = u = -\frac{B}{A} \quad\dots\dots\dots\dots\dots\text{(ii)}.$$

These values give the positions of the Principal Foci.

56. If P and P' be a pair of conjugate points upon the axis, we have

$$FP.F'P' = (AP - AF)(A'P' - A'F')$$

$$= \left(u + \frac{B}{A}\right)\left(v - \frac{C}{A}\right)$$

$$= \frac{Au + B}{A} \cdot \left(\frac{Cu + D}{Au + B} - \frac{C}{A}\right)$$

$$= \frac{AD - BC}{A^2}$$

$$= -\frac{1}{A^2} \quad\dots\dots\dots\dots\dots\dots\dots\dots\text{(i)}.$$

Hence FP and $F'P'$ are of contrary signs; that is, two points which are conjugate to one another must lie either both between or else both outside the Principal Foci.

It must be remembered that in this formula FP and $F'P'$ denote *reduced*, not *absolute* distances. The proper formula for absolute distances is

$$FP.F'P' = -\frac{\mu_0^2}{A^2} \quad\dots\dots\dots\dots\dots\text{(ii)}.$$

57. *The image formed by a lens.*

The image of a point is its conjugate point. Hence if an object be situated in a plane perpendicular to the axis, its image formed by a lens will also be in a plane perpendicular to the axis.

Also, if m_1 be the magnification produced by the first surface, m_2 that produced by the second surface, and m the resultant magnification produced by the lens, we clearly have

$$m = m_1 m_2.$$

We will now investigate formulæ for the determination of m.

58. *To investigate a formula for the magnification produced by a lens.*

In the case of a single refracting surface it has been proved that

$$m_1 = \frac{v_1}{u}$$

where u and v_1 are *reduced* distances.

Similarly for the second surface we have

$$m_2 = \frac{v}{v_1 + t};$$

therefore $\qquad \dfrac{1}{m_2} = \dfrac{v_1 + t}{v};$

But $\qquad \dfrac{1}{v} - \dfrac{1}{v_1 + t} = \sigma;$

therefore $\qquad \dfrac{v_1 + t}{v} = 1 + \sigma\,(v_1 + t);$

therefore $\qquad \dfrac{1}{m_1 m_2} = \dfrac{u}{v_1}\,[1 + \sigma\,(v_1 + t)]$

$$= \frac{u}{v_1}\,(1 + \sigma t) + u\sigma.$$

But $\qquad \dfrac{1}{v_1} - \dfrac{1}{u} = \rho;$

therefore $\qquad \dfrac{u}{v_1} = 1 + u\rho;$

therefore $\qquad \dfrac{1}{m_1 m_2} = (1 + u\rho)\,(1 + \sigma t) + u\sigma$

$$= 1 + \sigma t + u\,(\rho + \sigma + \rho\sigma t)$$

$$= B + Au;$$

therefore
$$\frac{1}{m} = B + Au \; \dots\dots\dots\dots\dots \; \text{(i)},$$

a formula which expresses the magnification in terms of u.

59. We may find a formula for m in terms of v in a similar way, or we may deduce it from Art. 58.

For, we have
$$v = \frac{Cu + D}{Au + B}.$$

Hence
$$u = \frac{D - Bv}{Av - C};$$

therefore
$$\frac{1}{m} = B + \frac{AD - ABv}{Av - C}$$
$$= \frac{AD - BC}{Av - C}$$
$$= \frac{1}{C - Av};$$

therefore
$$m = C - Av \; \dots\dots\dots\dots\dots \; \text{(ii)}.$$

60. *Helmholtz' Formula for the magnification.*

From Art. 38, we have for the first surface
$$m_1 = \frac{\mu_0 \tan \alpha}{\mu_1 \tan \alpha_1}.$$

Using a similar notation for the second surface, we get
$$m_2 = \frac{\mu_1 \tan \alpha_1}{\mu_2 \tan \alpha_2};$$

therefore
$$m = \frac{\mu_0 \tan \alpha}{\mu_1 \tan \alpha_1} \times \frac{\mu_1 \tan \alpha_1}{\mu_2 \tan \alpha_2},$$
$$= \frac{\mu_0 \tan \alpha}{\mu_2 \tan \alpha_2}.$$

61. *Points of unit magnification.*

If, in the formulæ for the magnification, we put $m = 1$, we get
$$\left. \begin{array}{l} u = \dfrac{1 - B}{A} \\[2mm] v = \dfrac{C - 1}{A} \end{array} \right\}.$$

These values of u and v determine two points on the axis, which we will denote by H and H' respectively. They are conjugate points; and are such, that if an object lie in· a plane through either of them perpendicular to the axis, an image of exactly the same size as the object will be produced on a corresponding plane through the other point.

62. These two points H and H' are of the very greatest importance in the discussion of the path of a ray of light through a thick lens, or through a system of thick lenses. They are fixed points, whose positions depend entirely upon the constants of the lens, and they may therefore be used very conveniently as origins, with reference to which the positions of other points may be reckoned.

Gauss was the first to introduce them into the problem. They were called by him *Haupt-puncte* or Principal Points. The planes through them, perpendicular to the axis, he called *Haupt-ebene* or Principal Planes.

We may define these Principal Planes and Principal Points as *Planes and Points of Unit Magnification.*

63. *Formulæ for the magnification, when the foci are the origins from which distances are measured.*

The foci are given by

$$u = -\frac{B}{A} \left.\vphantom{\frac{B}{A}}\right\} \cdot$$
$$v = \frac{C}{A} \left.\vphantom{\frac{C}{A}}\right.$$

Taking these points for origins, the magnification formulæ become

$$\frac{1}{m} = A\left(u - \frac{B}{A}\right) + B$$
$$= Au \dots\dots\dots\dots\dots\dots\dots (i),$$

and·

$$m = C - A\left(v + \frac{C}{A}\right)$$
$$= -Av \dots\dots\dots\dots\dots\dots\dots (ii),$$

u and v being measured from F and F' respectively.

Hence we see again that $uv = -\dfrac{1}{A^2}$ (Art. 56).

64. *Formulæ for the magnification when the Principal Points are the origins from which distances are measured.*

The principal points are given by

$$u = \frac{1 - B}{A} \left.\begin{matrix}\\\\\\\end{matrix}\right\}.$$
$$v = \frac{C - 1}{A}$$

Taking these points for origins, the magnification formulæ become

$$\frac{1}{m} = A\left(u + \frac{1 - B}{A}\right) + B$$

$$= Au + 1 \dots\dots\dots\dots\dots\dots\dots\text{(i)},$$

and
$$m = C - A\left(v + \frac{C - 1}{A}\right)$$

$$= -Av + 1 \dots\dots\dots\dots\dots\dots\text{(ii)},$$

u and v being measured from H and H' respectively.

65. *The formula connecting u and v when these distances are measured from the Principal Points.*

We have, from Art. 52, the relation

$$v = \frac{Cu + D}{Au + B},$$

u and v being measured from the *vertices*.

Hence, if we transfer the origins to the principal points, we get

$$v + \frac{C - 1}{A} = \frac{C\left(u + \dfrac{1 - B}{A}\right) + D}{A\left(u + \dfrac{1 - B}{A}\right) + B}$$

$$= \frac{ACu + C - BC + AD}{A^2 u + A - AB + BA}$$

$$= \frac{ACu + C - 1}{A(Au + 1)};$$

therefore $\quad (Av + C - 1)(Au + 1) = ACu + C - 1;$

therefore $\quad A^2uv + Av + Au(C-1) = ACu;$

therefore $\quad Auv + v - u = 0,$

or $\qquad\qquad \dfrac{1}{v} - \dfrac{1}{u} = A.$

66. If, in the formula,

$$\frac{1}{v} - \frac{1}{u} = A,$$

we put $v = \infty$, we get

$$u = -\frac{1}{A};$$

and if we put $u = \infty$, we get

$$v = \frac{1}{A}.$$

Hence $\qquad\qquad \left. \begin{array}{l} HF = -\dfrac{1}{A} \\[2mm] H'F' = \ \ \dfrac{1}{A} \end{array} \right\};$

that is, the reduced distance between either Focus and the corresponding Principal Point is equal to the reduced distance between the other Focus and the other Principal Point; and, moreover, the Foci are either both between or else both outside the Principal Points.

The *absolute* distances are $-\dfrac{\mu_0}{A}, \dfrac{\mu_0}{A}$, and are therefore also equal to one another, if the extreme media are the same; but they are not necessarily equal, if the extreme media are different.

67. *Definition.* The distance $H'F'$ is called the *Focal Length* of the Lens.

We may however take the *reduced* distance for the Focal Length, if it be distinctly understood that we do so. The two are the same when $\mu_0 = 1$.

If we denote it by f, we have $f = \dfrac{1}{A}$, and the formula of Art. 65 becomes $\dfrac{1}{v} - \dfrac{1}{u} = \dfrac{1}{f}$.

68. *Definition.* The quantity A is called the *Power* of the Lens.

69. The property of Principal Planes may be viewed in a slightly different way.

If we consider a pair of conjugate planes, and join any number of points on one plane to their conjugates on the other, we know that all these straight lines will meet the axis in one and the same point (Art. 46). But if one of the straight lines be parallel to the axis, it may be supposed to meet it at an infinite distance. Hence they will all meet the axis at an infinite distance, and consequently they will all be parallel to the axis.

But if two conjugate planes be so situated that the line joining any point on one of them to its conjugate on the other be parallel to the axis, it is obvious that to an object on one plane will correspond an image of exactly the same size on the other.

This is the property which belongs by definition to *Principal* Planes.

Consequently we may define Principal Planes and Principal Points as follows:

If two conjugate planes be such that the lines joining pairs of conjugate points on them are all parallel to the axis, these planes are called *Principal Planes*, and the points where they meet the axis are called *Principal Points*.

70. It may be noticed that conversely if any straight line parallel to the axis, meet the Principal Planes at the points P and P' respectively, then P and P' are conjugate to one another.

71. *To determine geometrically the position of a point P' which is conjugate to a given point P.*

We will assume that the Foci F and F', and the Principal Points H and H' have been determined, and that both the Principal Points lie between the Foci (fig. 17).

FIG. 17.

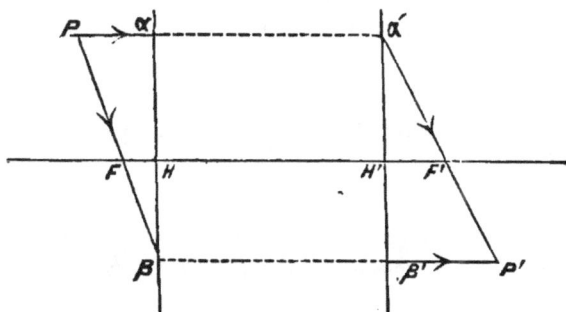

We know from the definition that all the rays from P will, after refraction, pass through P'; hence it will be sufficient to find the ultimate intersection of any two of them.

We will select that which proceeds from P in the direction $P\alpha$, parallel to the axis, and which meets the first Principal Plane at a point α; and also the ray $PF\beta$ which passes through the focus F and meets the Principal Plane at a point β.

The incident ray $P\alpha$ passes through I, the point at infinity, and α. Hence, (Art. 43), the corresponding emergent ray will pass through the points conjugate respectively to I and α. Let $P\alpha$ be produced to meet the other Principal Plane in α'. Then we know that α' is conjugate to α, and F' to I. Consequently $\alpha'F'$ will be the direction of the ray on emergence.

Again, the incident ray $PF\beta$ passes through F and β. If a straight line be drawn through β parallel to the axis, and meeting the other principal plane at β', then β' is conjugate to β. Also I' is conjugate to F. Hence $\beta'P'$ drawn parallel to the axis, will be the path of this ray on emergence.

The point P', which is conjugate to P, is the intersection of $\alpha'F'$ and $\beta\beta'$ produced.

We have consequently the following Geometrical construction for the determination of P'; through P draw $P\alpha\alpha'$ parallel to the axis to meet the second Principal Plane at α', and draw also $PF\beta$ to meet the first Principal Plane at β. Draw $\beta\beta'$ parallel to the axis. Then $\alpha'F'$ meets $\beta\beta'$, produced if necessary, at the point P' required.

72. It should be noticed that if the first and last media be the same, the figure $P\alpha'P'\beta$ will be a parallelogram.

73. *To investigate algebraical formulæ connecting the positions of two conjugate points which do not lie upon the axis.*

The letters in the accompanying figure have their customary signification.

We will assume, for simplicity, that H and H' lie between F and F', and we will consider numerical values only. The signs can be readily determined by inspection.

FIG.18.

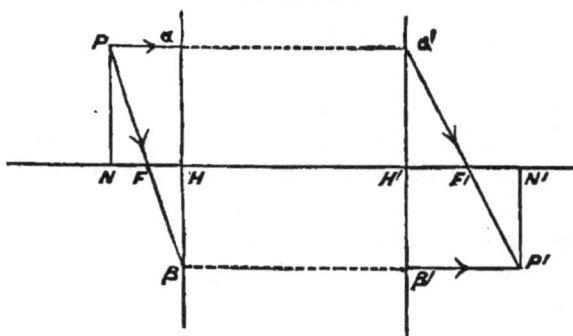

Let PN, $P'N'$ be drawn perpendicular to the axis; and let

$$\left. \begin{array}{ll} PN = y, & NH = x \\ P'N' = y', & N'H' = x' \\ HF = f, & H'F' = f' \end{array} \right\}.$$

Then, from the similar triangles βHF and $\beta\alpha P$, we get

$$\frac{HF}{\alpha P} = \frac{\beta H}{\beta\alpha},$$

or
$$\frac{f}{x} = \frac{y'}{y+y'}.$$

Again, from the triangles $\alpha' H' F'$ and $\alpha'\beta' P'$, we get in a similar way

$$\frac{f'}{x'} = \frac{y}{y+y'};$$

therefore
$$\frac{f}{x} + \frac{f'}{x'} = 1 \ \dots\dots\dots\dots\dots \ \text{(i)},$$

and
$$\frac{fy}{x} = \frac{f'y'}{x'} \ \dots\dots\dots\dots\dots \ \text{(ii)}.$$

If x and y be given, these two equations are sufficient to determine x' and y'. Hence, if P be given, the position of P' can be determined.

74. If the extreme media be the same, we have f and f' equal to one another numerically, but of opposite signs. In this case the formulæ of the preceding article reduce to

$$\left.\begin{array}{c} \dfrac{1}{x} + \dfrac{1}{x'} = \dfrac{1}{f} \\[2mm] \dfrac{y}{x} = \dfrac{y'}{x'} \end{array}\right\}.$$

and

The first of these is similar to the formula obtained in the case of refraction at one surface only; the distances u and v in that case being measured from the vertex.

The second formula, $\dfrac{y}{x} = \dfrac{y'}{x'}$, shows that if P and P' (fig. 19) be any two conjugate points, and if they be joined to H and H' respectively, the straight lines PH and $P'H'$ are parallel to one another; that is to say, an incident ray through the

FIG.19.

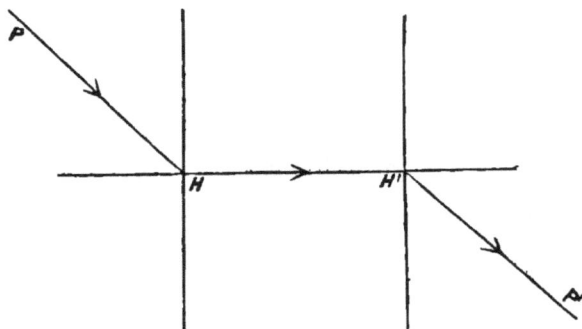

Principal Point H will produce a parallel emergent ray through the other Principal Point H'.

Hence for a lens, when the extreme media are the same, the Principal Points possess the property which belongs in the general case to what are called the *Nodal Points.* The general case will be considered in a subsequent chapter.

75. The results of Art. 73 may be put in another form. We have

$$\frac{x}{f} = \frac{y + y'}{y'};$$

therefore

$$\frac{x - f}{f} = \frac{y}{y'};$$

similarly

$$\frac{x' - f'}{f'} = \frac{y'}{y};$$

therefore

$$(x - f)(x' - f') = ff' \quad \ldots\ldots\ldots\ldots\ldots \text{(iii)}.$$

76. *If straight lines be drawn through F and F' perpendicular to the axis, to meet the incident and emergent rays respectively in D and D' then*

$$FD + F'D' = \Pi a.$$

Let the incident and emergent rays, produced if necessary, meet the axis at the points X, X' respectively (fig. 20).

FIG.20.

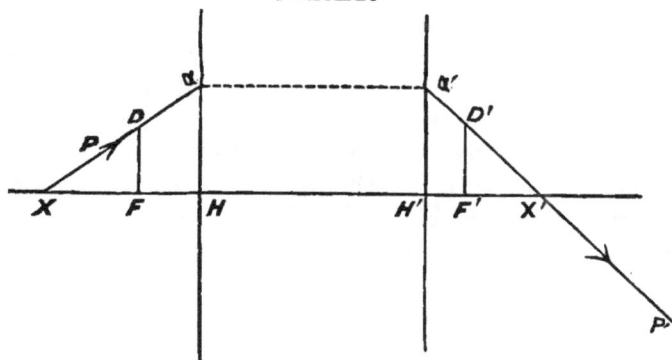

Then we know that X and X' are conjugate to one another; therefore

$$\frac{f}{HX} + \frac{f'}{H'X'} = 1;$$

therefore $\quad \dfrac{HX - f}{HX} + \dfrac{H'X' - f'}{H'X'} = 1;$

therefore $\quad \dfrac{FX}{HX} + \dfrac{F'X'}{H'X'} = 1;$

therefore $\quad \dfrac{FD}{H\alpha} + \dfrac{F'D'}{H'\alpha'} = 1;$

therefore $\quad FD + F'D' = H\alpha,$

for $\alpha\alpha'$ is parallel to the axis, and therefore $H'\alpha' = H\alpha$.

If we denote FD, $F'D'$, $H\alpha$ by z, z', h respectively, we get

$$z + z' = h,$$

which is similar to the corresponding formula obtained in Art. 16 in the case of a ray refracted at one surface only.[*]

77. *The image of a plane luminous object formed by a thick lens.*

We assume that the object lies on a plane P perpendicular to the axis; and that it may be regarded as a cluster of luminous points, each of which has its image on the plane

---

* Carl Neumann: *Ueber die Haupt- und Brenn- Puncte.*

P' conjugate to P. These point-images, in the aggregate form the image required.

The determination of the position of the plane P' when the Principal Planes have been found, has been explained in Arts. 71, 73. It will be shown in Chapter V. how the Principal Planes themselves may be determined experimentally.

Again, it has been proved that all the lines joining corresponding points of object and image, meet one another at a point C' on the axis. Hence the image is similar in form to the object, and will be inverted or upright according as C' lies between them or not.

78. *To determine the image graphically.*

Let us take any point of the object and join it to its conjugate point. This line, produced if necessary, will meet the axis at a point C'.

If we now describe a cone which has its vertex at C', and the object P for its base, then the section of this cone made by the conjugate plane P' will be the image required.

79. *Definition.* When a ray parallel to the axis is refracted by a lens, it receives a certain deviation. A *thin* lens, which, when placed coaxally with the lens, would produce the same deviation in the same ray, is said to be *equivalent* to the given lens. It is called, briefly, the *Equivalent Lens.*

80. *To find the deviation produced by a thick lens.*

Let $QPP'Q'$ be the course of a ray, which before incidence *is parallel to the axis*, and which crosses the refracting surfaces at the points P and P' respectively (fig. 20 a).

Let $A_1P = h$, and let the radii of the surfaces be r and s.

We will consider all deviations positive when they are *towards* the axis, and we will denote the deviation at P by $-\delta$; then from Art. 8 we have

$$-\delta = \frac{\mu_1 - \mu_0}{\mu_1} \frac{h}{r}$$

$$= \frac{\rho h}{\mu_1} \quad\dots\dots\dots\dots\dots\dots\dots\dots\dots \text{(i)}.$$

FIG.20.(a)

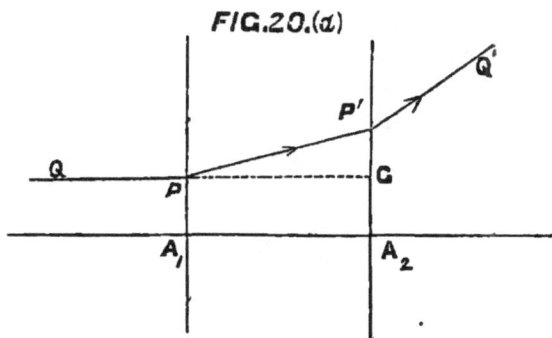

We have, from the Geometry of the figure,

$$A_2 P' = A_1 P + PG \tan P'PG$$
$$= h + t\delta$$
$$= h + t\frac{\rho h}{\mu_1}, \text{ if } -t = \text{absolute thickness of lens,}$$
$$= h(1 + \rho t), \text{ if } -t = \text{reduced thickness.}$$

Again, deviation at P'

$$= \frac{\mu_0 - \mu_1}{\mu_0}\left\{\frac{A_2 P'}{s} - \delta\right\}, \text{ by Art. 8,}$$
$$= \frac{\mu_0 - \mu_1}{\mu_0}\left\{\frac{h(1 + \rho t)}{s} - \delta\right\}$$
$$= \frac{h\sigma(1 + \rho t)}{\mu_0} - \frac{\mu_0 - \mu_1}{\mu_0}\delta.$$

But total deviation = deviation at P + deviation at P';

therefore total deviation $= +\delta + \dfrac{h\sigma(1 + \rho t)}{\mu_0} - \dfrac{\mu_0 - \mu_1}{\mu_0}\delta$

$$= \frac{h\sigma(1 + \rho t)}{\mu_0} + \frac{\mu_1}{\mu_0}\delta$$
$$= \frac{h}{\mu_0}\{\sigma(1 + \rho t) + \rho\}$$
$$= \frac{Ah}{\mu_0}.$$

E

81. *The focal length of the equivalent Lens.*

In Art. 9 we have shewn that a single refracting surface, which may be considered as an indefinitely thin lens, produces in a ray parallel to the axis a deviation

$$= \frac{h}{r} \frac{\mu_1 - \mu_0}{\mu_1}$$

$$= \frac{h}{f},$$

f being the *absolute* focal length of the surface.

Also (Art. 80) a thick lens produces a deviation

$$= \frac{Ah}{\mu_0}.$$

Hence the absolute focal length of the thin lens which would produce the same deviation as a given thick lens, or, in other words, the absolute focal length of the *equivalent lens*, is given by the formula

$$\frac{h}{f} = \frac{Ah}{\mu_0},$$

or $f = \dfrac{\mu_0}{A}.$

Consequently, the *reduced* focal length of the equivalent lens

$$= \frac{1}{A}.$$

82. If the results of Arts. 67 and 81 be compared, it will be seen that the focal length of the equivalent lens is equal to what we have defined as the focal length of the thick lens itself, for each is equal to $\dfrac{1}{A}$.

This equality is evident geometrically. For if we consider a ray which before incidence is parallel to the axis, we know that on emergence it will pass through F' (see fig. 21); the deviation being represented in the figure by the angle between $P\alpha\alpha'$ produced and $\alpha'F'$. And if the ray pass through the equivalent lens, the deviation produced being the same,

FIG. 21.

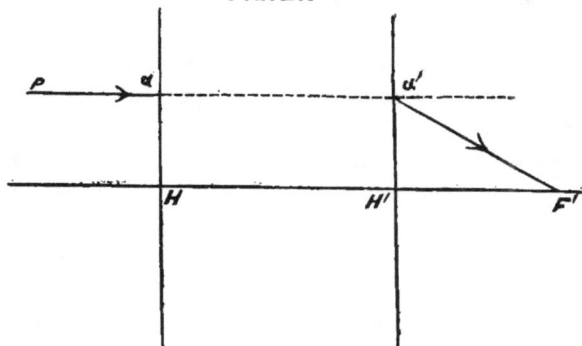

it is clear that the emergent ray will cut the axis at a distance from the lens equal to $H'F'$, the same as before. Consequently the two focal lengths are equal to one another.

83. In this chapter we have considered the passage of a ray of light across *two* surfaces only, and we have proved (Arts. 52, 58, 59) that the positions of conjugate points are connected by the relation

$$v = \frac{Cu + D}{Au + B},$$

and that the *magnification* is given by the formula

$$m = C - Av,$$

or

$$\frac{1}{m} = B + Au.$$

In the next chapter we shall prove that however many surfaces be crossed by the ray, the formulæ which correspond to those given above are exactly analogous to them; and that in the general case, for any number of surfaces, the constants A, B, C, D are connected with one another in exactly the same manner as in the case of an ordinary thick lens (Art. 53).

CHAPTER III.

84. The refracting surfaces to be considered are spherical, and have all their centres of curvature upon the same straight line, which is the axis of the system.

The spaces between the surfaces are supposed to be occupied by homogeneous media, such that the medium on one side of any surface and that on the other side have different refractive indices.

By assigning suitable values to the indices of refraction, to the distances between the vertices, and to the radii of curvature, this system of refracting surfaces may be adapted to the case of any system whatever of any number of co-axal lenses, simple or compound.

85. Several of the properties that have been proved for a lens can be extended at once to the general case. It will be sufficient if we merely state the propositions, and then leave the reader to prove them by generalising the corresponding propositions in Chapter II. In many cases the requisite alterations will be but verbal.

86. (i) *If any number of rays proceed from a luminous point P, and traverse a system of any number of refracting surfaces in succession, they will, after emergence from the system, pass through one and the same point P′* (Art. 41).

Two points such as P and $P′$ are said to be conjugate to one another with respect to the system of surfaces.

(ii) *The point of concurrence of any two incident rays is conjugate to the point of concurrence of the corresponding emergent rays.*

(iii) *If P, P' be a pair of conjugate points, and p, p' another pair, then the incident ray which passes through P and p will produce an emergent ray passing through P' and p'.*

(iv) *To a system of points P lying on a plane perpendicular to the axis, corresponds a system of points P', which are conjugate respectively to the points P, and also lie on a plane perpendicular to the axis.*

Two such planes are called *Conjugate Planes*, that is to say, planes conjugate to one another with respect to the system of surfaces considered. The points where they meet the axis are called *Conjugate Foci*.

(v) *If P and P' be two conjugate points, a plane through P perpendicular to the axis will be conjugate to a plane through P' perpendicular to the axis.*

(vi) *If two conjugate planes be taken, and any number of points on one be joined to their conjugates on the other, all these straight lines will meet the axis at the same point.*

87. If the incident rays are all parallel to the axis, they may be considered as proceeding from a point I on the axis at an infinite distance from the vertex of the first surface. After emergence from the system they will meet together at a point F' on the axis.

Again, the rays which after emergence from the system are all parallel to the axis may be supposed to meet the axis at an infinitely distant point I', and must before incidence on the first surface have proceeded from a certain point F on the axis.

The points F and F' are called the *Principal Foci*, or briefly, *the* Foci of the system. The planes through them perpendicular to the axis are the *Focal Planes* of the system.

Also F, I' and I, F' are pairs of conjugate points.

The properties (Art. 50) of rays which before incidence or after emergence, meet at a point on a Focal Plane, are true also whatever be the number of surfaces crossed by the rays.

88. We will now suppose that the source of light is at a point X on the axis of the system, and we will denote by u

its *reduced* distance from the vertex of the refracting surface nearest to it. The reduced distance of X_1, the point conjugate to X with respect to the first surface, will be denoted by v_1.

If X_2 be conjugate to X_1 with respect to the second surface, their distances from the second vertex may be denoted by

$$v_1 + t_1 \text{ and } v_2,$$

where, with the notation explained in Art. 52, $-t_1$ is the reduced distance between the two vertices.

Again, if $-t_2$ be the distance between the second and third vertices, the distances from the third vertex of X_2 and the point conjugate to it with respect to the third surface may be denoted by

$$v_2 + t_2 \text{ and } v_3;$$

and so on.

In this way, if we suppose that there are n surfaces, we have for the last pair of distances

$$v_{n-1} + t_{n-1} \text{ and } v_n.$$

We will suppose the radii of curvature of the surfaces to be $r_1, r_2, r_3, \ldots r_n$ successively; and the indices of refraction of the successive media to be $\mu_0, \mu_1, \mu_2, \mu_3, \ldots \mu_n$; μ_n and μ_0 being equal to one another, when the first and last media are the same.

The expressions

$$\frac{\mu_1 - \mu_0}{r_1}, \; \frac{\mu_2 - \mu_1}{r_2}, \; \ldots \; \frac{\mu_n - \mu_{n-1}}{r_n},$$

will be denoted by $\rho_1, \rho_2, \rho_3, \ldots \rho_n$ respectively.

The above notation will be employed throughout the subsequent articles. It will be observed that the distances are *reduced* distances.

We will now proceed to consider the relation between the positions of two conjugate foci.

89. *To find the relation between the positions of two conjugate foci, when a ray is refracted at n surfaces in succession.*

From Art. 52 we have, with the notation explained in the preceding article,

$$\frac{1}{v_1} - \frac{1}{u} = \rho_1,$$

$$\frac{1}{v_2} - \frac{1}{v_1 + t_1} = \rho_2,$$

$$\frac{1}{v_3} - \frac{1}{v_2 + t_2} = \rho_3,$$

$$\cdots\cdots\cdots\cdots\cdots\cdots$$

$$\cdots\cdots\cdots\cdots\cdots\cdots$$

$$\frac{1}{v_n} - \frac{1}{v_{n-1} + t_{n-1}} = \rho_n.$$

Hence

$$v_n = \cfrac{1}{\rho_n + \cfrac{1}{t_{n-1} + v_{n-1}}}$$

$$= \frac{1}{\rho_n +} \frac{1}{t_{n-1} +} \frac{1}{\rho_{n-1} +} \frac{1}{t_{n-2} +} \cdots \frac{1}{t_1 +} \frac{1}{\rho_1 +} \frac{1}{u}.$$

90. If $\dfrac{C}{A}$ and $\dfrac{D}{B}$ be the penultimate and ante-penultimate convergents to the continued fraction

$$\frac{1}{\rho_n +} \frac{1}{t_{n-1} +} \frac{1}{\rho_{n-1} +} \cdots \frac{1}{t_1 +} \frac{1}{\rho_1 +} \frac{1}{u},$$

we get

$$v_n = \frac{Cu + D}{Au + B}.$$

Now, the number of quantities such as

$$u, \rho_1, t_1, \rho_2, t_2, \cdots\cdots t_{n-1}, \rho_n$$

is $2n$.

Hence $\dfrac{C}{A}$ is an odd, and $\dfrac{D}{B}$ an even convergent to v_n; consequently

$$\frac{D}{B} - \frac{C}{A} = -\frac{1}{BA};$$

therefore

$$AD - BC = -1.$$

For simplicity we will suppress the subscript letter in v_n and call the distance v. We have then

$$v = \frac{Cu + D}{Au + B} \quad \text{...................... (i)},$$

where the constants A, B, C and D are connected by the relation

$$AD - BC = -1 \quad \text{..................... (ii)}.$$

It will be seen that the formulæ (i) and (ii) are precisely analogous to those obtained in Arts. 52 and 54, in the case of a single thick lens.

91. If in the formula

$$v = \frac{Cu + D}{Au + B}$$

we put $u = \infty$, we get

$$v = \frac{C}{A} \quad \text{.......................... (i)},$$

and if we put $v = \infty$, we get

$$u = -\frac{A}{B} \quad \text{.......................... (ii)},$$

which determine the positions of the *Principal Foci* of the system.

92. If P, P' be two conjugate points on the axis, and F, F'' be the Foci, we have, as in Art. 56,

$$FP.F'P' = -\frac{1}{A^2},$$

if FP and $F'P'$ denote *reduced* distances, or

$$FP.F'P' = -\frac{\mu_0 \mu_n}{A^2},$$

if they denote *absolute* distances.

93. In Art. 53, it was shewn that each of the constants B, C, D can be expressed in terms of A. This is a very important property, and we will now show that it is a general one; that it holds for the case of a system of n surfaces as well as for a simple thick lens.

94. If $\dfrac{1}{a+}\ \dfrac{1}{b+}\ \dfrac{1}{c+...}$ denote any continued fraction, its successive convergents may be written in the form

$$\dfrac{(1)}{(a)},\ \dfrac{(b)}{(ab)},\ \dfrac{(bc)}{(abc)},\ \dfrac{(bcd)}{(abcd)},\ \&c.$$

the successive numerators and denominators being connected by the relations

$$(bc) = c\,(b) + (1),$$
$$(bcd) = d\,(bc) + (b),$$
$$(abcd) = d\,(abc) + (ab),$$
$$\&c.$$

The functional sign is purposely the same throughout, for we know that (bcd) in the fourth numerator is exactly the same function of b, c, d, that (abc) in the third denominator is of a, b, c.

Among the known properties of the numerators and denominators of these successive convergents we have the following;

$$(abcd...hk) \equiv (kh...dcba) \ .... \text{ I,}$$

that is to say, any function $(abcd...hk)$ is unaltered if we reverse the order of the letters. Also

$$\left.\begin{array}{l} (bc) = \dfrac{d}{da}\,(abc) = \dfrac{d}{da}\,(cba) \\[2mm] (bcd) = \dfrac{d}{da}\,(abcd) = \dfrac{d}{da}\,(dcba) \\[2mm] \qquad\qquad \&c. \end{array}\right\} \ .... \text{ II.}$$

These theorems are very important, but we have not been able to find them in any English book on the subject. It may therefore be useful to the reader if we give proofs of them in an Appendix at the end of this volume.

95. The quantity A is the denominator of the penultimate convergent to the continued fraction

$$\dfrac{1}{\rho_n+}\ \dfrac{1}{t_{n-1}+}\ \dfrac{1}{\rho_{n-1}+}\ ......\ \dfrac{1}{t_1+}\ \dfrac{1}{\rho_1+}\ \dfrac{1}{u}\,,$$

and is consequently a function of

$$\rho_n, \ t_{n-1}, \ \rho_{n-1}, \ \cdots \ t_1, \ \rho_1.$$

Hence, we may write

$$A = \phi \ (\rho_n, \ t_{n-1}, \ \rho_{n-1}, \ \cdots \ t_1, \ \rho_1),$$

and therefore by Theorem I of the preceding article, we have also

$$A = \phi \ (\rho_1, \ t_1, \ \rho_2, \ t_2, \ \cdots \ \rho_{n-1}, \ t_{n-1}, \ \rho_n) \cdots\cdots\cdots (\text{i}) \ ;$$

and, by Theorem II, we have

$$\left. \begin{array}{l} B = \dfrac{dA}{d\rho_1} \\[2mm] C = \dfrac{dA}{d\rho_n} \\[2mm] D = \dfrac{d^2 A}{d\rho_1 \, d\rho_n} \end{array} \right\} \cdots\cdots\cdots\cdots (\text{ii}).$$

Hence, whatever be the number of surfaces, B, C and D can always be expressed in terms of A.

These results are precisely analogous to those obtained in Art. 53.

96. Let us represent the denominators of the successive convergents to the continued fraction

$$\frac{1}{\rho_n +} \ \frac{1}{t_{n-1} +} \ \frac{1}{\rho_{n-1} +} \ \cdots\cdots \ \frac{1}{t_1 +} \ \frac{1}{\rho_1},$$

by

$$A_1, \ A_2, \ A_3, \ \cdots\cdots \ A_{2n-1} \ ;$$

then we know that each of these quantities is connected with the two preceding it by the relations

$$A_{2n-1} - \rho_1 A_{2n-2} - \ A_{2n-3} \qquad\qquad = 0,$$
$$A_{2n-2} - t_1 A_{2n-3} - \ A_{2n-4} \qquad\qquad = 0,$$
$$A_{2n-3} - \rho_2 A_{2n-4} - A_{2n-5} = 0,$$
$$\cdots\cdots\cdots\cdots\cdots\cdots\cdots \ \&\text{c.,}$$

of which the last are

$$A_3 - t_{n-1} A_2 - \ A_1 = 0,$$
$$A_2 - \rho_n A_1 = 1.$$

97. If, for example, we suppose that there are *four* surfaces, we have $2n - 1 = 7$, and we get

$$
\left.
\begin{aligned}
A_7 - \rho_1 A_6 - A_5 &= 0 \\
A_6 - t_1 A_5 - A_4 &= 0 \\
A_5 - \rho_2 A_4 - A_3 &= 0 \\
A_4 - t_2 A_3 - A_2 &= 0 \\
A_3 - \rho_3 A_2 - A_1 &= 0 \\
A_2 - t_3 A_1 &= 1 \\
A_1 &= \rho_4
\end{aligned}
\right\}.
$$

Hence, solving these equations for A_7, we get

$$
A_7 \begin{vmatrix}
1, & -\rho_1, & -1, & 0, & 0, & 0, & 0 \\
0, & 1, & -t_1, & -1, & 0, & 0, & 0 \\
0, & 0, & 1, & -\rho_2, & -1, & 0, & 0 \\
0, & 0, & 0, & 1, & -t_2, & -1, & 0 \\
0, & 0, & 0, & 0, & 1, & -\rho_3, & -1 \\
0, & 0, & 0, & 0, & 0, & 1, & -t_3 \\
0, & 0, & 0, & 0, & 0, & 0, & 1
\end{vmatrix}
$$

$$
= \begin{vmatrix}
0, & -\rho_1, & -1, & 0, & 0, & 0, & 0 \\
0, & 1, & -t_1, & -1, & 0, & 0, & 0 \\
0, & 0, & 1, & -\rho_2, & -1, & 0, & 0 \\
0, & 0, & 0, & 1, & -t_2, & -1, & 0 \\
0, & 0, & 0, & 0, & 1, & -\rho_3, & -1 \\
1, & 0, & 0, & 0, & 0, & 1, & -t_3 \\
\rho_4, & 0, & 0, & 0, & 0, & 0, & 1
\end{vmatrix}.
$$

The coefficient of A_7 in the above result is a determinant which has all the terms on one side of its diagonal zero, and all the terms in the diagonal unity. The value of the determinant reduces therefore to unity; and we get

$$
A_7 = - \begin{vmatrix}
-\rho_1, & -1, & 0, & 0, & 0, & 0, & 0 \\
1, & -t_1, & 1, & 0, & 0, & 0, & 0 \\
0, & 1, & -\rho_2, & -1, & 0, & 0, & 0 \\
0, & 0, & 1, & -t_2, & -1, & 0, & 0 \\
0, & 0, & 0, & 1, & -\rho_3, & -1, & 0 \\
0, & 0, & 0, & 0, & 1, & -t_3, & -1 \\
0, & 0, & 0, & 0, & 0, & 1, & -\rho_4
\end{vmatrix}.
$$

98. If, again, we suppose that there are *two* refracting surfaces, we have $2n - 1 = 3$, and therefore

$$\left. \begin{array}{c} A_3 - \rho_1 A_2 - A_1 = 0 \\ A_2 - t_1 A_1 = 1 \\ A_1 = \rho_2 \end{array} \right\};$$

therefore
$$A_3 = - \left| \begin{array}{ccc} -\rho_1, & -1, & 0 \\ 1, & -t_1, & -1 \\ 0, & 1, & -\rho_2 \end{array} \right|$$

$$= \rho_1 \rho_2 t_1 + \rho_1 + \rho_2,$$

which agrees with the result obtained for a thick lens in the preceding chapter.

99. For the case of *one* surface only, we have

$$A_1 = - (- \rho_1)$$

$$= \rho_1,$$

which, too, agrees with previous results.

100. For the general case of n surfaces, we have

$$A_{2n-1} = - \left| \begin{array}{cccccc} -\rho_1, & -1, & 0, & 0, & 0, & \dots\dots\dots\dots \\ 1, & -t_1, & -1, & 0, & 0, & \dots\dots\dots\dots \\ 0, & 1, & -\rho_2, & -1, & 0, & \dots\dots\dots\dots \\ 0, & 0, & 1, & -t_2, & -1, & \dots\dots\dots\dots \\ \multicolumn{6}{c}{\dots\dots\dots\dots\dots\dots\dots\dots\dots\dots\dots\dots} \\ \multicolumn{6}{c}{\dots\dots\dots\dots\dots\dots\dots 1, \; -t_{n-1}, \; -1} \\ \multicolumn{6}{c}{\dots\dots\dots\dots\dots\dots\dots 1, \quad -\rho_n} \end{array} \right| .$$

101. We may now investigate a relation whereby the value of A for n surfaces may be determined from that for $n - 1$ surfaces; that is to say, we will investigate an equation connecting A_{2n-1} and A_{2n-3}.

We know that A_{2n-1} is the denominator of the last convergent to the continued fraction

$$\frac{1}{\rho_n +} \frac{1}{t_{n-1} +} \frac{1}{\rho_{n-1} +} \frac{1}{t_{n-2} +} \dots \frac{1}{t_1 +} \frac{1}{\rho_1},$$

that is, to the continued fraction

$$\frac{1}{\rho_1}+\frac{1}{t_1}+\frac{1}{\rho_2}+\frac{1}{t_2}+\cdots\frac{1}{t_{n-1}}+\frac{1}{\rho_n},$$

obtained by reversing the order of the letters in the preceding one.

If we denote the denominators of the successive convergents to the latter fraction by

$$A_1, A_2, A_3, \ldots A_{2n-1},$$

we have

$$A_{2n-1} = \rho_n A_{2n-2} + A_{2n-3} \\ A_{2n-2} = t_{n-1} A_{2n-3} + A_{2n-4} \Big\};$$

therefore

$$A_{2n-1} = (\rho_n t_{n-1} + 1) A_{2n-3} + \rho_n A_{2n-4}$$

$$= (\rho_n t_{n-1} + 1) A_{2n-3} + \rho_n \frac{dA_{2n-3}}{d\rho_{n-1}}.$$

102. It is clear that

$$A_1 = \rho_1 \quad\ldots\ldots\ldots\ldots\ldots\ldots\ldots\text{(i)}.$$

From this, by means of the formula just proved, we get

$$A_3 = (1 + \rho_2 t_1) A_1 + \rho_2 \frac{dA_1}{d\rho_1}$$

$$= (1 + \rho_2 t_1) \rho_1 + \rho_2$$

$$= \rho_1 + \rho_2 + \rho_1 \rho_2 t_1 \quad\ldots\ldots\ldots\ldots\ldots\text{(ii)}.$$

Hence

$$A_5 = (1 + \rho_3 t_2) A_3 + \rho_3 \frac{dA_3}{d\rho_2}$$

$$= (1 + \rho_3 t_2)(\rho_1 + \rho_2 + \rho_1 \rho_2 t_1) + \rho_3 (1 + \rho_1 t_1)\ldots\text{(iii)},$$

and so on.

The formula of Art. 101 consequently enables us to determine the value of A for three surfaces from its value for two, then its value for four surfaces from its value for three, and thus by successive steps to its value for any number of surfaces whatever. Moreover, from the value of A in any particular case, we can determine the corresponding values of B, C and D. Thus we see that the values of A, B, C, D for any system of surfaces may be deduced from the value of A for *one* surface only.

103. *Magnification.* It has been proved in the particular case of a lens, that m or the magnification can be expressed as a linear function of v, and $\dfrac{1}{m}$ as a linear function of u.

We will now show that the magnification can be so expressed, whatever be the number of refracting surfaces; that the formulæ in the general case are precisely analogous to those obtained in the particular case; and that they involve the constants A, B, and C in exactly the same way.

104. *The magnification produced by a system of n refracting surfaces can be expressed as a linear function of v.*

Let us consider the system formed by the first $n-1$ surfaces, and let $(m)_{n-1}$ denote the magnification produced by it.

Also let v_{n-1} be the reduced distance of the $(n-1)^{\text{th}}$ image from the $(n-1)^{\text{th}}$ vertex.

We will assume that $(m)_{n-1}$ can be expressed as a linear function of v_{n-1}, and thence show that on this supposition, the magnification can again be expressed as a linear function of v_n, if we cause the rays to pass through an additional or n^{th} surface; v_n being the reduced distance of the nth image from the vertex of the nth surface.

We will assume that

$$(m)_{n-1} = \gamma - a v_{n-1},$$

and we will suppose that $-t_{n-1}$ is the *reduced* distance between the $(n-1)^{\text{th}}$ and the nth vertices.

It has been proved, that the magnification produced by a single surface

$$= \frac{v}{u}.$$

Consequently the magnification caused by the nth surface alone

$$= \frac{v_n}{v_{n-1} + t_{n-1}}.$$

Hence, if we denote by m the whole magnification produced by the n surfaces, we have

$$m = (\gamma - \alpha v_{n-1}) \frac{v_n}{v_{n-1} + t_{n-1}} .$$

But

$$\frac{1}{v_n} - \frac{1}{v_{n-1} + t_{n-1}} = \rho_n ;$$

therefore

$$\frac{v_n}{v_{n-1} + t_{n-1}} = 1 - v_n \rho_n ;$$

therefore

$$m = \{\gamma + \alpha t_{n-1} - \alpha (v_{n-1} + t_{n-1})\} \frac{v_n}{v_{n-1} + t_{n-1}}$$

$$= \frac{v_n}{v_{n-1} + t_{n-1}} (\gamma + \alpha t_{n-1}) - \alpha v_n$$

$$= (1 - v_n \rho_n) (\gamma + \alpha t_{n-1}) - \alpha v_n$$

$$= \gamma + \alpha t_{n-1} - v_n \{\alpha + \rho_n (\gamma + \alpha t_{n-1})\}$$

$$= \gamma' - \alpha' v_n.$$

Hence the magnification can be expressed as a linear function of v_n.

105. Again, if P and P' be two points that are conjugate to one another with respect to the system of surfaces, we know that which ever we consider as the source of light, the rays from it will produce an image at the other point. The same is true of objects and images of a definite size. Consequently, if we transpose the terms object and image in the preceding article, and consider as an origin of light what is there treated as an image, the image of it produced by the system of surfaces will be what we originally considered as the object.

The magnification produced by the system thus transposed will clearly

$$= \frac{1}{m} .$$

Also the quantity corresponding to v_n is obviously u.

Hence, we obtain as a result, that for a system of any

number of surfaces the inverse magnification, or $\dfrac{1}{m}$, can be expressed as a linear function of u, or

$$\frac{1}{m} = \beta' + \alpha''u.$$

106. We have proved so far that the magnification produced by a system of n refracting surfaces may be expressed in either of the forms

$$m = \gamma' - \alpha'v_n,$$

and

$$\frac{1}{m} = \beta' + \alpha''u.$$

If we suppress the subscript letter n, the former becomes

$$m = \gamma' - \alpha'v.$$

These formulæ are similar in form to those obtained for a lens. We have now to determine the yet unknown constants β', α'', γ', α'.

107. The formulæ for the magnification, namely,

$$\left.\begin{array}{c} m = \gamma' - \alpha'v \\ \dfrac{1}{m} = \beta' + \alpha''u \end{array}\right\},$$

must clearly be equivalent; so that if in the former we were to put for v its known value in terms of u, we should certainly get the latter. It follows, therefore, that if we eliminate m from the two expressions, the result of elimination will be an expression involving only u, v, and constants, which must be identically the same as the known relation

$$v = \frac{Cu + D}{Au + B}.$$

Eliminating m, we get

$$1 = (\gamma' - \alpha'v)(\beta' + \alpha''u),$$

or

$$1 = \gamma'\beta' + \alpha''\gamma'u - \beta'\alpha'v - \alpha'\alpha''uv,$$

or

$$\alpha'\alpha''uv + \beta'\alpha'v - \alpha''\gamma'u + 1 - \gamma'\beta' = 0 \quad\ldots\ldots\ldots \text{(i)}$$

Also from $v = \dfrac{Cu + D}{Au + B}$, we get

$$Auv + Bv - Cu - D = 0 \dots\dots\dots\dots (ii).$$

The equations (i) and (ii) must be identically the same; hence, comparing coefficients, we have

$$\frac{\alpha'\alpha''}{A} = \frac{\beta'\alpha'}{B} = \frac{\alpha''\gamma'}{C} = \frac{1 - \gamma'\beta'}{-D};$$

therefore
$$\left.\begin{aligned} \frac{\alpha'}{A} &= \frac{\gamma'}{C} = \lambda \\[2mm] \frac{\alpha''}{A} &= \frac{\beta'}{B} = \lambda' \end{aligned}\right\} \dots\dots\dots\dots (iii);$$

therefore, substituting for α', α'', γ', β', in the equation

$$\frac{\alpha'\alpha''}{A} = \frac{1 - \gamma'\beta'}{-D},$$

we get $\qquad \lambda\lambda'AD = -1 + \lambda\lambda'BC;$

therefore $\qquad \lambda\lambda'(AD - BC) = -1;$

therefore $\qquad \lambda\lambda' = 1 \dots\dots\dots\dots\dots (iv).$

Again, substituting in the formulæ for m the values of α', α'', β', γ' given by (iii), we get

$$m = \lambda (C - Av),$$
$$\frac{1}{m} = \frac{1}{\lambda} (B + Au).$$

The resemblance between the formulæ for n surfaces and the formulæ for a lens has now become more distinct; but the method described here does not lead to the determination of λ in an elementary way. The method is in itself important, but for our special purpose it will be better to consider the matter directly.

108. *To investigate a formula for the magnification produced by a system of n refracting surfaces.*

Let us consider the last $n - 1$ surfaces and let A', B', C', D' be the corresponding values of A, B, C, D.

F

If m' denote the magnification produced by these $n-1$ surfaces, we will assume that

$$\frac{1}{m'} = A'u' + B',$$

where u' is the distance of the second image from the second vertex.

The magnification produced by the first surface is given by

$$\frac{1}{m_1} = \frac{u}{v_1}.$$

Hence m, or the resultant magnification produced by the whole system, is given by

$$\frac{1}{m} = (A'u' + B')\frac{u}{v_1}$$

$$= \{A'(v_1 + t_1) + B'\}\frac{u}{v_1}$$

$$= A'u + (B' + A't_1)\frac{u}{v_1}.$$

But $$\frac{1}{v_1} - \frac{1}{u} = \rho_1;$$

therefore $$\frac{u}{v_1} = 1 + u\rho_1;$$

therefore $$\frac{1}{m} = A'u + (B' + A't_1)(1 + u\rho_1)$$

$$= u(A' + B'\rho_1 + A't_1\rho_1) + B' + A't_1.$$

Now A' is the denominator of the last convergent to the continued fraction

$$\frac{1}{\rho_2 +}\ \frac{1}{t_2 +}\ \frac{1}{\rho_3 +} \ldots \frac{1}{\rho_n};$$

we have also $$B' = \frac{dA'}{d\rho_2},$$

and we have proved that successive denominators are connected with one another by the equation

$$A_{2n-1} = (\rho_n t_{n-1} + 1) A_{2n-3} + \rho_n \frac{dA_{2n-3}}{d\rho_{n-1}}, \quad \text{(Art. 101).}$$

Hence $$A = (\rho_1 t_1 + 1) A' + \rho_1 B',$$

and $$B = \frac{dA}{d\rho_1} = A't_1 + B';$$

therefore

$$\frac{1}{m} = Au + B \dots\dots\dots\dots\dots\dots\dots\text{(i)} ;$$

and if in this formula we substitute for u its value in terms of v found from the equation

$$v = \frac{Au + B}{Cu + D},$$

we get

$$m = -Av + C \dots\dots\dots\dots\dots \text{(ii)}.$$

Hence the magnification produced by a system of n refracting surfaces can be expressed in exactly the same form as that produced by a lens.

109. If the foci be taken as the origins of distances, the formulæ for the magnification become (Art. 63)

$$\left.\begin{aligned} \frac{1}{m} &= Au \\ m &= -Av \end{aligned}\right\},$$

u and v being measured from F and F' respectively.

110. *Helmholtz' formula for the magnification produced by a system of n surfaces.*

If $\mu_0, \mu_1, \mu_2, \dots \mu_n$ be the refractive indices of the successive media;

and $\alpha, \alpha_1, \alpha_2, \dots \alpha_n$ be the angles at which the successive portions of the ray are inclined to the axis of the system;

and $m_1, m_2, m_3, \dots m_n$ the magnifications produced by the surfaces; we have

$$m_1 = \frac{\mu_0 \tan \alpha}{\mu_1 \tan \alpha_1},$$

$$m_2 = \frac{\mu_1 \tan \alpha_1}{\mu_2 \tan \alpha_2},$$

$$\dots\dots\dots\dots\dots,$$

$$m_n = \frac{\mu_{n-1} \tan \alpha_{n-1}}{\mu_n \tan \alpha_n};$$

therefore

$$m = \frac{\mu_0 \tan \alpha}{\mu_n \tan \alpha_n}.$$

111. *Points of Unit Magnification.*

If in the formulæ for the magnification we put $m = 1$, we get

$$u = \frac{1 - B}{A} \\ v = \frac{C - 1}{A} \Bigg\} .$$

The points so determined were called by Gauss the *Principal Points* of the system of surfaces; the planes through them perpendicular to the axis are the *Principal Planes.*

The Principal Planes are conjugate to one another with respect to the system of surfaces, and are such that to an object on one plane will correspond an image on the other, the image and the object being of exactly the same size.

112. From the definition of Principal planes, that they are Conjugate Planes and also Planes of Unit Magnification, it follows that a straight line parallel to the axis will meet them in points that are conjugate to one another.

113. All that we have said with respect to the Principal Points of a lens is equally true of the Principal Points of a system of n surfaces.

It will be enough perhaps if we simply state the facts, the proofs of them being already given in the corresponding articles of Chapter II.

(i) If the Principal Points be taken as origins from which distances are measured, we get

$$m = - Av + 1 \\ \frac{1}{m} = Au + 1 \Bigg\} , \qquad \text{(Art. 64),}$$

$$\frac{1}{v} - \frac{1}{u} = A, \qquad \text{(Art. 65).}$$

(ii)
$$HF = -\frac{1}{A} \Bigg\}, \quad \text{(Art. 66)},$$
$$H'F' = \frac{1}{A}$$

these being *reduced* distances.   In *absolute* measurements

$$HF = -\frac{\mu_0}{A} \Bigg\}.$$
$$H'F' = \frac{\mu_n}{A}$$

If the two extreme media be the same, these absolute distances become numerically equal.

The absolute distance $H'F'$ is called the *Focal Length* of the system.

If we denote the *reduced* distance $H'F'$ by f, we get $f = \frac{1}{A}$, and therefore

$$\frac{1}{v} - \frac{1}{u} = \frac{1}{f}.$$

Also the Foci lie either both between or both beyond the Principal Points.

(iii) To determine the position of the point P' conjugate to a given point P, we have the following construction:— Through P draw $P\alpha\alpha'$ parallel to the axis to meet the second Principal Plane at α', and draw also $PF\beta$ to meet the first Principal Plane at β.   Draw $\beta\beta'$ parallel to the axis.   Then $\alpha'F'$ produced will meet $\beta\beta'$ at the point P' required. (Art. 71).

(iv) If (xy), $(x'y')$ be the coordinates of two conjugate points P and P' respectively, as explained in Art. 73, we have

$$\frac{f}{x} + \frac{f'}{x'} = 1 \Bigg\},$$
$$\frac{fy}{x} = \frac{f'y'}{x'}$$

f and f' being the *absolute* distances HF, $H'F'$ respectively.

Also
$$(x - f)(x' - f') = ff'.$$

(v) If straight lines be drawn through F and F' perpendicular to the axis to meet the incident ray and the emergent ray respectively in D and D', then

$$FD + F'D' = H\alpha,$$

α being the point at which the incident ray meets the Principal Plane. (Art. 76).

114. The image produced by a system of surfaces can be determined in the same manner as in the case of a lens. (Arts. 77, 78).

115. *Definition.* When a ray parallel to the axis is refracted by a system of surfaces, the ray receives a certain deviation.

A *thin* lens, which when placed coaxally with the system would produce the same deviation in the same ray, is said to be *equivalent* to the given system of surfaces.

It is called, briefly, the *Equivalent Lens.*

116. *To find the deviation produced by a system of surfaces.*

If a ray parallel to the axis cross a refracting surface at a distance h from the axis, it has been shown (Art. 80) that the deviation produced

$$= \frac{\rho h}{\mu},$$

where μ is the refractive index of the medium following the surface, and ρ the power of the surface.

Let the deviations after crossing the successive surfaces be

$$\delta_1, \ \delta_2, \ \delta_3, \ \dots \ \delta_n;$$

and the distances from the axis of the points of incidence be

$$h_1, \ h_2, \ h_3, \ \dots \ h_n;$$

the powers of the surfaces being as before

$$\rho_1, \ \rho_2, \ \rho_3, \ \dots \ \rho_n.$$

Assuming as before that the deviations are measured towards the axis we get easily (as in Art. 80) that,

$$\delta_1 = \frac{h_1 \rho_1}{\mu_1},$$

$$h_2 = h_1 + t_1 \delta_1,$$

$$\delta_2 = \frac{\mu_1 \delta_1 + h_2 \rho_2}{\mu_2},$$

and so on, $-t_1$, &c. being the *absolute* thicknesses.

Now if we write δ'_1 for $\mu_1 \delta_1$, δ'_2 for $\mu_2 \delta_2$, &c., these equations become, using *reduced* thicknesses,

$$\delta'_1 = h_1 \rho_1,$$

$$h_2 = h_1 + \delta'_1 t_1,$$

$$\delta'_2 = \delta'_1 + h_2 \rho_2,$$

$$h_3 = h_2 + \delta'_2 t_2,$$

$$\cdots\cdots\cdots\cdots$$

$$\cdots\cdots\cdots\cdots$$

$$\delta'_n = \delta'_{n-1} + h_n \rho_n.$$

If we form the continued fraction

$$\frac{1}{\rho_1 +} \ \frac{1}{t_1 +} \ \frac{1}{\rho_2 +} \ \frac{1}{t_2 +} \ \&c. + \frac{1}{\rho_n},$$

it is easy to see that the denominator of the last convergent will be $\dfrac{\delta'_n}{h_1}$, *i.e.* $\dfrac{\mu_n \delta_n}{h_1}$; and the same denominator has also been represented by A. Hence

$$\delta_n = \frac{h_1 A}{\mu_n}.$$

117. If f be the absolute focal length of the equivalent lens, we have

$$\frac{h_1}{f} = \delta_n;$$

therefore

$$f = \frac{h_1}{\delta_n} = \frac{\mu_n}{A},$$

therefore the reduced focal length $= \dfrac{1}{A}$.

118. Again, since the denominators of the last convergents to the fractions

$$\frac{1}{\rho_n +} \ \frac{1}{t_{n-1} +} \ \frac{1}{\rho_{n-1} +} \ \cdots \ \frac{1}{\rho_1}$$

and

$$\frac{1}{\rho_1 +} \ \frac{1}{t_1 +} \ \frac{1}{\rho_2 +} \ \cdots \ \frac{1}{\rho_n}$$

are the same, it follows that the two reduced focal lengths of the system are equal to one another.

Hence if we denote the absolute lengths by f and f', we have

$$\frac{f}{\mu_0} = \frac{f'}{\mu_n}.$$

If the first and last media are the same, we have $\mu_0 = \mu_n$; hence

$$f = \ f' \quad \text{numerically,}$$

or
$$\cdot f = - f' \quad \text{algebraically.}$$

119. It should be noticed that the so-called equivalent lens merely produces *the same deviation* as the system. It does not bring the rays to the same focus, nor therefore does it produce an image in the same position. It may be shown however that both these conditions may be satisfied by using *two* thin lenses properly situated.

CHAPTER IV.

ACHROMATISM.

120. When an object is viewed through a lens or through a system of lenses it commonly appears to have a sort of coloured border. This is due to the fact that ordinary sunlight is composite and not simple. It is a combination of an indefinitely large number of·different kinds of light, which have different degrees of refrangibility and all varieties of colour.

The various component rays which make up a resultant ray of ordinary sun-light can be separated from one another by allowing the ray to pass through a glass prism. If the light as it emerges be cast upon a screen, there will appear an elongated continuous coloured band. The component rays being of different refrangibilities must necessarily meet the screen at different points.

Now, if P be a luminous point and P' be its conjugate point, or image, we know that the position of P' is a function of the refractive indices of a ray with respect to the various media. Consequently each component of the light that proceeds from P will produce a separate image, in a separate position, and of its own proper colour. Moreover, the magnification also is a function of the refractive indices, and therefore the images formed by the component lights will all be of different sizes.

These separate images are all formed on planes perpendicular to the axis, and one behind another. The image actually seen by the eye is the resultant of them, that is to say, the image obtained by their superposition. The central portion of it is colourless as ordinary light, for there every component plays its part. But towards the edge, in consequence of the difference in the sizes and positions of the component images, one colour after another ceases to appear and the resultant image is seen surrounded with a rainbow-like border.

So long as the image has this coloured edge, it is indefinite and unsatisfactory. We will endeavour to show whether it can be got rid of, and if it can, by what means and under what conditions.

A lens, or system of lenses, which produces an image with a distinct border free from colour is said to be *achromatic*.

121. The problem before us is simply this:—If an object be viewed through a lens or system of lenses, what conditions must the lens or lenses satisfy in order that the images formed by rays of two or more colours may be coincident.

It is clear from what has been said that (i) the focus-conjugate to a given point on the axis must be the same for two or more colours, and (ii) the magnification must be the same for two or more colours.

It should be remarked that the refractive indices differ but slightly from one another and that they all lie within certain limits.

122. We will now express the conditions in an algebraical form, by means of the equations

$$\frac{1}{m} = Au + B,$$

$$v = \frac{Cu + D}{Au + B}.$$

The distances u, v being *reduced* depend on the distances refractive indices of the first and last media. In most Astronomical instruments these are both the same, and vary very little (under ordinary circumstances) for different colours. We shall therefore suppose u to be unaltered by the variation of the refractive index of the first medium, and v to vary only by reason of the indices contained in A, B, C, D.

123. Considering the second requirement, we have

$$\frac{1}{m} = Au + B;$$

therefore $\qquad \delta \left(\frac{1}{m}\right) = u\delta A + \delta B,$

u being constant, and the symbol δ expressing variation due to variations in the values of the μ's.

If the magnification be the same for all, then

$$\delta \left(\frac{1}{m}\right) = 0;$$

therefore $\qquad u\delta A + \delta B = 0.$

This must be true for all values of u,

therefore $\qquad\qquad \delta A = 0$
and $\qquad\qquad\qquad \delta B = 0$ $\quad$ .... (i).

These are the conditions that the different coloured images may be all of the same size.

124. We will now consider the other requirement, that the images be formed all in the same place.

We have

$$v = \frac{Cu + D}{Au + B},$$

and we have to find the condition that v may remain the same while A, B, C, and D undergo small variations. This condition gives us

$$\delta \left(\frac{Cu + D}{Au + B}\right) = 0.$$

But, from Art. 123,

$$Au + B = \text{const.};$$

therefore $\qquad \delta (Cu + D) = 0;$

therefore $\qquad u\delta C + \delta D = 0.$

This must be true for all values of u;

therefore $\qquad\qquad \delta C = 0$
$\qquad\qquad\qquad\qquad \delta D = 0$ $\quad$ .... (ii).

These are the conditions that the component images may be all in the same plane.

125. If we combine the results of Arts. 123 and 124 we see that both requirements will be satisfied if

$$\delta A = \delta B = \delta C = \delta D = 0 \dots\dots\dots\dots(iii).$$

The quantities A, B, C, D however are connected by the relation

$$AD - BC = -1,$$

and the four conditions contained in (iii) are not all independent.

The necessary and sufficient conditions for achromatism are

$$\delta A = \delta B = \delta C = 0,$$

or $$\delta A = \delta C = \delta D = 0,$$

or $$\delta B = \delta C = \delta D = 0.$$

126. If it be required that the system shall be achromatic only for a particular value of u, we must eliminate u between the two equations

$$\left.\begin{array}{l} u\delta A + \delta B = 0 \\ u\delta C + \delta D = 0 \end{array}\right\};$$

therefore the sufficient condition is

$$\frac{\delta D}{\delta C} = \frac{\delta B}{\delta A} = -u.$$

127. In Art. 177 and in subsequent Articles of Parkinson's *Optics* the problem of achromatism is considered in two ways, according as the pencil passes centrically or excentrically. These methods are important in such an instrument, for example, as the telescope.

128. The former case may be considered by the methods which we have just explained.

When the pencil passes centrically through the system, all the conjugate foci lie upon the axis, and our object is gained if we make as many as possible of these coincide.

The condition is

$$\delta v = 0.$$

But $$v = \frac{Cu + D}{Au + B},$$

therefore $(u\delta A + \delta B)(Cu + D) = (Au + B)(u\delta C + \delta D);$

therefore $u^{2}(A.\delta C - C.\delta A) + u(A.\delta D + B.\delta C - D.\delta A - C.\delta B)$

$$+ B.\delta D - D.\delta B \qquad = 0.$$

129. The case of excentrical refraction can best be considered by the method given in Parkinson's *Optics*, Art. 180. We may however obtain a similar result by making

$$\delta A' = 0,$$

where A' is the power of the quasi-equivalent lens, that is, the lens which would produce the same deviation as the system does in a ray initially *inclined* and not parallel to the axis. An expression for A' may be obtained in a way similar to that in Art. 117.

130. The quantities δA, δB, &c., may be determined as follows. We have

$$A = f(\rho_{1}, t_{1}, \rho_{2}, t_{2} \cdots \rho_{n}),$$

where $-t_{1}$, $-t_{2}$, ... denote the *reduced* distances between the vertices of the surfaces; therefore

$$\delta A = \frac{\delta A}{\delta \rho_{1}} \delta \rho_{1} + \frac{\delta A}{\delta t_{1}} \delta t_{1} + \ldots + \frac{\delta A}{\delta \rho_{n}} \delta \rho_{n}.$$

Also $$\rho_{1} = \frac{\mu_{1} - \mu_{0}}{r_{1}},$$

$$t_{1} = \frac{t'_{1}}{\mu_{1}},$$

&c.,

where $-t'_{1}$, $-t'_{2}$, ... are the *absolute* distances between the vertices.

Therefore $\quad \delta\rho_p = \dfrac{1}{r_p} \delta(\mu_p - \mu_{p-1})$

$$= \dfrac{1}{r_p}(\delta\mu_p - \delta\mu_{p-1});$$

$$\delta t_p = \delta\left(\dfrac{t'_p}{\mu_p}\right)$$

$$= -t'_p \dfrac{\delta\mu_p}{\mu^2_p}$$

$$= -t_p \dfrac{\delta\mu_p}{\mu_p}.$$

In this way we may get $\delta\rho_1$, δt_1, &c.; hence, by substitution, we get δA in terms of $\delta\mu_1$, ..., &c.

131. Considering a single thick lens of refractive index μ and thickness $-a$, and supposing the external air to be of constant index 1, we have

$$A = \rho_1 + \rho_2 + \rho_1\rho_2 t,$$

where $\quad \rho_1 = \dfrac{\mu-1}{r_1}, \quad \rho_2 = \dfrac{1-\mu}{r_2}, \quad t = \dfrac{a}{\mu}.$

Hence $\quad \delta A = (1 + \rho\, t)\,\delta\rho_1 + (1 + \rho_1 t)\,\delta\rho_2 + \rho_1\rho_2\delta_1,$

and $\quad \delta\rho_1 = \dfrac{d\mu}{r_1}, \quad \delta\rho_2 = -\dfrac{\delta\mu}{r_2}, \quad \delta t = -\dfrac{a\delta\mu}{\mu^2}.$

The condition $\delta A = 0$ gives us

$$\left(1 - \dfrac{\mu-1}{r_2}\dfrac{a}{\mu}\right)\dfrac{\delta\mu}{r_1} - \left(1 + \dfrac{\mu-1}{r_1}\dfrac{a}{\mu}\right)\dfrac{\delta\mu}{r_2}$$

$$+ \dfrac{(\mu-1)^2}{r_1 r_2}\dfrac{a\delta\mu}{\mu^2} = 0;$$

or $\quad \dfrac{1}{r_1} - \dfrac{1}{r_2} - \dfrac{\mu-1}{\mu}\dfrac{a}{r_1 r_2}\left\{2 - \left(\dfrac{\mu-1}{\mu}\right)\right\} = 0 \dots\dots(i).$

Again $\qquad\qquad B = 1 + \rho_2 t.$

Calculating the variation of B in the same manner, and making it equal to zero, we get either $a = 0$, or $\dfrac{1}{r_2} = 0$, or $\dfrac{1}{\mu} = 0$.

In the first case the two radii must be equal in order to satisfy (i), and the lens ceases to be of any optical value. The second case $\frac{1}{r_2} = 0$ needs $\frac{1}{r_1}$ to be zero also, with the same result as before. Lastly, $\frac{1}{\mu} = 0$ is an impossibility. Our conclusion therefore is that a single lens cannot be achromatic.

CHAPTER V.

132. *To determine theoretically the positions of the Foci
and of the Principal Points of a lens or system of lenses.*

Conjugate Foci lying upon the axis are connected with
one another by the relation

$$\frac{f}{x} + \frac{f'}{x'} = 1 \dots\dots\dots\dots\text{(i)},$$

where x, x' are the distances of the conjugate points from
the Principal Points H, H' respectively.

Let us take any point O on the axis of the system,
and use it as an origin from which all our distances may
be reckoned; and let F, F', H, H' be the distances from
it of the Principal Foci and the Principal Points respectively.

The above relation may now be written in the form

$$\frac{H - F}{H - \xi} + \frac{H' - F'}{H' - \xi'} = 1 \dots\dots\dots\dots\text{(ii)},$$

where ξ, ξ' are the distances from O of any pair of conjugate points.

We have shown, in Art. 113, how to determine the
position of the point conjugate to a given one. If therefore
we take any four points on the axis, and determine the
positions of the four points respectively conjugate to them,
we get four simultaneous values of ξ and ξ'. These, when
substituted successively in (ii), give us *four* independent
equations for the determination of the *four* unknown quantities F, F', H, H'. Thus are found the positions of the
Principal Foci and also of the Principal Points.

133. *To determine experimentally the positions of the Foci of a lens or system of lenses.*

We will suppose that, in the following figure, *A* is a micrometer or a frame holding two spider lines crossing one another, *C* a stand supporting a telescope, and that *B* supports a cylinder enclosing the system of lenses. Moreover *A*, *B*, and *C* are supposed to be moveable by the hand or by means of screws to and fro along the graduated bar *MN*, and also to be so adjusted that the micrometer, the lenses, and the telescope have the same axis (fig. 22).

FIG. 22.

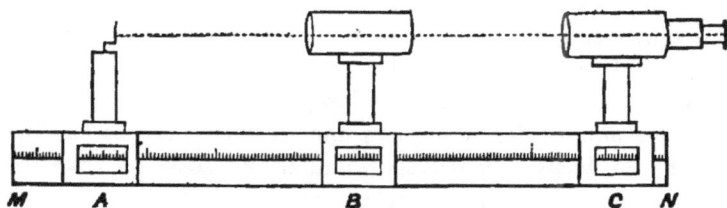

Let the telescope be turned first towards a *distant* object, and then accurately focused. The rays from the distant object are approximately parallel, and the image will be formed at the Principal Focus of the telescope.

When this has been done, the telescope must be placed on the stand *C* so that the micrometer *A* may be viewed through the system of lenses; the micrometer must then be moved to and fro along the graduated bar until the image of it, seen through the system of lenses and the telescope, becomes clear and distinct. Now this image is seen through a telescope which has been focused upon a distant object, hence we know that the image of the micrometer can be distinct only when the rays that fall upon the object glass of the telescope are parallel to the axis. Consequently the rays that emerge from the cylinder *B* must be parallel to the axis. Therefore the micrometer *A* must be at one of the Principal Foci of the lens-system. The Focus is thus determined in position.

G

We have still to measure its distance from the nearest surface of the system. This might be done by moving the micrometer along the graduated bar until it came into contact with the surface, and then taking the difference of the readings in the two positions given by the scale. There is however a practical difficulty in ascertaining the exact moment of contact, and this method consequently leads to an unsatisfactory result. The distance between the micrometer and the nearest surface may be measured more accurately by a simple optical contrivance.

For this purpose let the telescope be focused upon some near object whose distance is greater than that which we have to measure; and let the telescope be removed to the other end of the bar so that the micrometer may be between it and the lens-system. If when this has been done the telescope be moved along the graduated bar until first the micrometer, and then the *dust on the face of the lens* be in focus successively; and if the scale be read for these two positions of the telescope, the difference between the readings will give us the distance between the micrometer and the face of the lens-system with tolerable accuracy. The micrometer being at a focus of the system, we thus get the distance of the focus from the surface nearest to it.

134. *To determine the positions of the Principal Points when the Foci are known.*

If d and d' be the distances from the Foci of any two conjugate points on the axis, and if f be the distance of a Focus from the corresponding Principal Point, we have

$$dd' = f^2.$$

It has been shown how d and d' may be determined; hence the above equation enables us to determine f, and consequently the positions of H and H'.

135. In order that the method described above may be a directly practical one, it is necessary that the lens-system should be a convex lens or equivalent to a

convex lens. Otherwise no real images will be formed by it.

If the lens-system itself gives a real image, the method can be applied at once. But if it does not, it can be made to do so by combining with it a known convex lens of sufficient power. The method may then be applied to the joint-system, and by making allowance for the effect of the known convex lens, the proper result for the original system can be deduced from the one thus obtained.

136. The Principal Points, introduced by Gauss, have been supplemented by two other points which Listing intro-duced and called *Nodal Points*. They are principally of importance when the extreme media are not the same. This is found to be the case in the .human eye.

The Nodal Points are situated upon the axis of the lens-system, and are conjugate to one another. Their dis-tinguishing property is that an incident ray through one will produce an emergent ray in a parallel direction passing through the other.

When the extreme media are the same, we have seen that this is a property of the Principal Points. Hence in this case the Principal Points and the Nodal Points coincide.

When there is only *one* refracting surface we might call its centre of curvature the Nodal Point, for we know that an incident ray which passes through the centre of curvature crosses the surface without deviation.

137. *To determine the positions of the Nodal Points.*

Let H, H', F, F'' be the Principal Points and Foci of a lens-system, and let T be any point on the Focal Plane through F (fig. 23).

We know that a ray through T parallel to the axis, and meeting the Principal Planes at α, α' respectively, will on emergence pass through F'. Moreover, since T is a point in a focal plane, its conjugate is on a plane at an infinite distance, and therefore all rays from it will on

FIG.23.

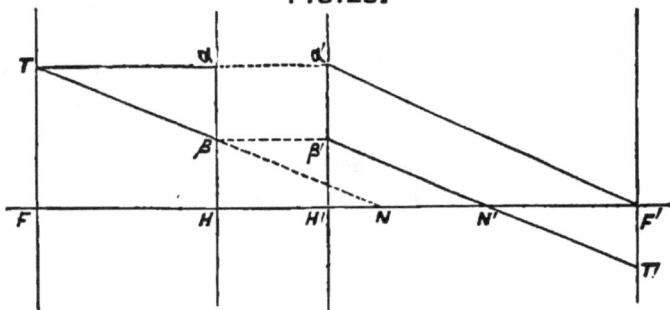

emergence be parallel to one another. They will therefore all be parallel to $\alpha'F'$.

Hence if $T\beta$ meet the first Principal Plane at β, and if $\beta\beta'$ drawn parallel to the axis meet the other Principal Plane at β', then $\beta'N'$ drawn parallel to $\alpha'F'$ will be the emergent ray produced by the incident ray $T\beta$.

Let $T\beta$ produced and the corresponding emergent ray meet the axis of the system at the points N, N' respectively.

Then we see from the figure that the triangles TFN and $\alpha'H'F'$ are equal in all respects; therefore

$$FN = H'F' = \text{constant}.$$

Hence the position of N is independent of the position of T; therefore N is a fixed point, and its distance from F is equal to the second focal distance.

In a similar way it may be shown that

$$F'N' = FH;$$

therefore N' also is a fixed point independent of T, and is at a distance from F' equal to the first focal distance.

The points N and N' are clearly the Nodal Points referred to in Art. 136.

138. If the extreme media be the same, the two focal distances are equal; hence, as we have already noticed, the points N and N' coincide with H and H'.

139. From the figure we have also

$$NN' = \beta\beta' = HH'.$$

☞ Hence the distance between the Nodal Points is equal to the distance between the Principal Points.

140. We have also

$$HN = H'N' = H'F' - HF.$$

141. When the Nodal Points have been determined, we may with their help very readily determine the direction of the emergent ray produced by a given incident ray, and also the position of a point conjugate to a given one (fig. 24).

FIG. 24.

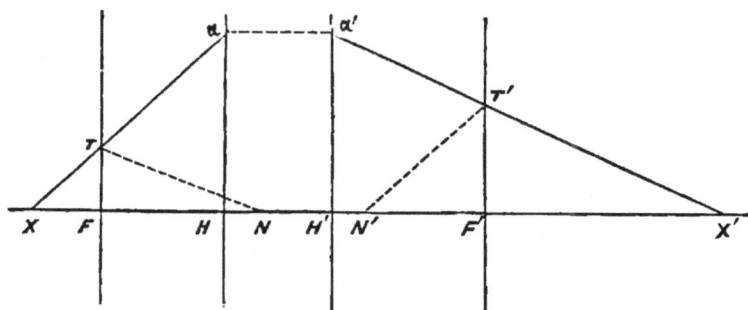

With the usual notation we will suppose $XT\alpha$ to be an incident ray meeting the Focal Plane through F at T. Join TN. Let $\alpha\alpha'$ parallel to the axis meet the second Principal Plane at α'. Through α' draw $\alpha'T'X'$ parallel to TN. Then $\alpha'T'X'$ is the direction of the emergent ray produced by the incident ray $XT\alpha$, and X, X' are a pair of conjugate points.

Otherwise: Draw $N'T''$ parallel to $XT\alpha$ and meeting the Focal plane through F' at the point T ; then $\alpha'T''$ is the direction of the emergent ray.

If an object lens be situated on the Nodal Plane through N its image will be situated on the Nodal Plane through N', and it may easily be proved that the linear dimensions of object and image are to one another inversely as the indices of refraction of the first and last media.

CHAPTER VI.

142. WE will now apply some of the results of Chapter II to determine the positions of the Principal Points and Foci for the five most important forms of the simple lens.

We will consider, (1) a double convex lens, (2) a plano-convex lens, (3) a double concave lens, (4) a plano-concave lens, (5) a meniscus; and finally we will consider (6) the case of two mirrors placed upon the same axis, and facing one another, in such manner as we find in Gregory's and Cassegrain's telescopes.

143. In Arts. 55 and 64, it was shown that the distances of the Foci and of the Principal Points from the vertices of the lens are given by the formulæ

$$AF = -\frac{B}{A} = -\frac{t\sigma + 1}{\rho\sigma t + \rho + \sigma},$$

$$A'F' = \frac{C}{A} = \frac{t\rho + 1}{\rho\sigma t + \rho + \sigma},$$

$$AH = \frac{1-B}{A} = -\frac{t\sigma}{\rho\sigma t + \rho + \sigma},$$

$$A'H' = \frac{C-1}{A} = \frac{t\rho}{\rho\sigma t + \rho + \sigma}.$$

We will now suppose the refractive index of the outside medium to be *unity*, and that of the substance of the lens to be μ. The values of ρ and σ become therefore $\frac{\mu - 1}{r}$ and $\frac{1 - \mu}{s}$ respectively.

Also for the *reduced* thickness $-t$ we will snbstitute its value, $-\dfrac{t}{\mu}$, in terms of the *absolute* thickness.

When the expressions have been simplified we find that

$$AF = \frac{-\mu rs + (\mu - 1)\, tr}{(\mu - 1)\,\{\mu\,(s - r) - (\mu - 1)\, t\}}\,,$$

$$A'F'' = \frac{\mu rs + (\mu - 1)\, ts}{(\mu - 1)\,\{\mu\,(s - r) - (\mu - 1)\, t\}}\,,$$

$$AH = \frac{tr}{\mu\,(s - r) - (\mu - 1)\, t}\,,$$

$$A'H' = \frac{ts}{\mu\,(s - r) - (\mu - 1)\, t}\,;$$

and consequently

$$HF = \frac{-\mu rs}{(\mu - 1)\,\{\mu\,(s - r) - (\mu - 1)\, t\}}\,.$$

These formulæ belong to such a lens as is described in fig. 14, which we have chosen for our standard. To obtain the corresponding formulæ for any one of the particular forms, we have merely to make the proper changes in the signs of the radii r and s.

We shall assume, and the assumption is practically correct, that r and s are both greater than the thickness of the lens.

We may now proceed to consider the six cases in turn.

144. *Case I. A double convex lens.*

In this case s is negative, and we must therefore change the sign of s in the formulæ given above.

We see then, $-t$ being a positive quantity, that AH is positive and $A'H'$ negative, and that both are numerically less than the thickness of the lens. Consequently both the Principal Points are situated within the material of the lens.

Also, since the sum of the distances AH and $A'H'$ considered numerically is found to be greater than the thickness, it follows that the points A, A', H, H', are disposed in the order

$$A, \quad H', \quad H, \quad A'.$$

The focal length of the lens is negative, therefore the focal length of the equivalent lens is negative. Consequently a double-convex lens is a convergent one; that is to say, the deviation of any ray passing through it will be in the direction of the axis.

145. *Case II. A Plano-convex lens.*

In this case one of the radii, r suppose, is infinite; and the other s is negative. The formulæ therefore become

$$AF = \frac{-\mu s - (\mu - 1)\, t}{\mu\, (\mu - 1)},$$

$$A'F' = \frac{s}{\mu - 1},$$

$$AH = -\frac{t}{\mu},$$

$$A'H' = 0.$$

The Principal Points are therefore situated, one at the vertex of the curved surface, and the other in the interior of the lens.

The focal length is given by

$$HF = \frac{-s}{\mu - 1},$$

and this is also the focal length of the equivalent lens. Hence the lens is convergent.

146. *Case III. A double concave lens.*

In this case r is negative. Hence we see that AH is positive and less than the thickness; so that the point H lies within the substance of the lens. Similarly H' also must lie within the lens.

Again $AH + A'H'$ is numerically less than t, therefore the points A, A', H, H' are disposed in the order

$$A, \quad H, \quad H', \quad A'.$$

Also we see from the formulæ that AF is positive and $A'F'$ negative, so that F lies on the side of the lens farther from A, or in other words behind the lens, and the point F' in front of the lens.

We should notice too, that in this case the points F and F' are only *virtual* foci, whereas in the case of a double-convex lens the foci are *real*.

Lastly, the focal length of the lens itself, and therefore of the equivalent lens, is positive. Therefore the lens is divergent.

147. *Case IV. A plano-concave lens.*

This case may be treated in the same way as Case II. We must put $r = \infty$, s being positive.

We then find that one Principal Plane is at the vertex of the curved surface, and the other within the lens. Also that the foci are virtual, and the lens divergent.

148. *Case V. A meniscus.*

This form of lens is represented in fig. 14, which is our standard form. The corresponding formulæ are those given in Art. 143.

We see therefore that the focal length will be positive or negative according as

$$r - s \gtrless \frac{(\mu - 1)}{\mu} \times \text{thickness}.$$

This is also the condition that, for parallel rays, the foci may be virtual or real respectively.

For the case of the meniscus, which has its concavity turned the other way, we must change the signs of both r and s in the standard formulæ.

149. *Case VI. Two coaxal mirrors.*

We will suppose that two concave mirrors are situated upon the same axis, and face one another as in Gregory's telescope.

The positions of the fundamental points may be determined from our standard formulæ, by changing the sign of s, and putting $\mu = -1$.

Thus we get

$$AF = \frac{rs + 2tr}{2\,(r + s + 2t)},$$

$$A'F' = \frac{-rs - 2ts}{2\,(r + s + 2t)},$$

$$AH = \frac{tr}{r + s + 2t},$$

$$A'H' = \frac{-ts}{r + s + 2t},$$

and the focal length $\quad = -\tfrac{1}{2}\dfrac{rs}{r + s + 2t}.$

In the same way as before it may be shown that the Principal Points are disposed in the order

$$A, \quad H, \quad H', \quad A',$$

and that the equivalent lens is negative.

The system of mirrors in Cassegrain's telescope may be treated in the same manner.

APPENDIX.

PROPERTIES OF CONTINUED FRACTIONS.

1. THE property which we have quoted in Art. 94, namely, that the last denominator of the continued fraction

$$\frac{1}{\rho_1 +} \frac{1}{t_1 +} \frac{1}{\rho_2 +} \dots \frac{1}{\rho_n}$$

is equal to the last denominator of the continued fraction

$$\frac{1}{\rho_n +} \frac{1}{t_{n-1} +} \frac{1}{\rho_{n-1} +} \dots \frac{1}{\rho_1}$$

may be found in *Vorlesungen über Zahlen Theorie* von P. G. Lejeune Dirichlet.

2. Let a, b be any two quantities, and γ, δ, ε, ... λ, μ, ν a series of any other quantities.

Let us from these form another series

$$c, \ d, \ e, \ \dots \ l, \ m, \ n,$$

such that these latter quantities are connected with the former by the relations

$$\left.\begin{aligned}
c &= \gamma b + a \\
d &= \delta c + b \\
e &= \varepsilon d + c \\
&\dots\dots\dots \\
n &= \nu m + l
\end{aligned}\right\} \dots\dots\dots\dots\dots(1).$$

From the first and second of the relations we get

$$\begin{aligned}
d &= \delta (\gamma b + a) + b \\
&= (\delta \gamma + 1) b + \delta a.
\end{aligned}$$

Substituting this value of d in the third we get

$$e = \varepsilon \{(\delta\gamma + 1) b + \delta a\} + \gamma b + a$$
$$= (\varepsilon\delta\gamma + \varepsilon + \gamma) b + (\varepsilon\delta + 1) a.$$

Hence, continuing the substitution, we get eventually

$$n = Ga + Hb \dots\dots\dots\dots\dots\dots\dots(2),$$

where G and H are functions of γ, δ, ε, ... λ, μ, ν, and are independent of a and b.

We will denote H by $\phi(\gamma, \delta, \varepsilon, ... \lambda, \mu, \nu)$, and we may then write (2) in the form

$$n = Ga + \phi(\gamma, \delta, \varepsilon, ... \lambda, \mu, \nu) b \dots\dots\dots\dots(3).$$

If we were to consider only the quantities $b, c, d, ... n$ in the one series, and δ, ε, ... λ, μ, ν in the other, we should in a similar way obtain the relation

$$n = G'b + \phi(\delta, \varepsilon, ... \lambda, \mu, \nu) c.$$

Substituting now $\gamma b + a$ for c, we get

$$n = G'b + (\gamma b + a) \phi(\delta, \varepsilon, ... \lambda, \mu, \nu)$$
$$= \phi(\delta, \varepsilon, ... \lambda, \mu, \nu) a + \{G' + \gamma.\phi(\delta, \varepsilon, ... \lambda, \mu, \nu)\} b \dots(4).$$

If the expressions for n in (3) and (4) be compared, it will be obvious that

$$G = \phi(\delta, \varepsilon, ... \lambda, \mu, \nu);$$

and therefore, by analogy,

$$G' = \phi(\varepsilon, ... \lambda, \mu, \nu).$$

From (3) and (4) we also get

$$H = G' + \gamma.\phi(\delta, \varepsilon, ... \lambda, \mu, \nu).$$

Hence, substituting in this equation the expressions already obtained for H and G', we get

$$\phi(\gamma, \delta, \varepsilon, ... \lambda, \mu, \nu) = \gamma.\phi(\delta, \varepsilon, ... \lambda, \mu, \nu) + \phi(\varepsilon, ... \lambda, \mu, \nu) \dots(5).$$

Again, if we put $a = 0$, $b = 1$, in the last three of the relations (1), we get

$$n = \phi (\gamma, \delta, \varepsilon, \dots \lambda, \mu, \nu),$$
$$m = \phi (\gamma, \delta, \varepsilon, \dots \lambda, \mu),$$
$$l = \phi (\gamma, \delta, \varepsilon, \dots \lambda);$$

and if we substitute these expressions for n, m, and l, in the relation

$$n = \nu m + l,$$

we get

$$\phi (\gamma, \delta, \varepsilon, \dots \lambda, \mu, \nu) = \nu.\phi (\gamma, \delta, \varepsilon, \dots \lambda, \mu) + \phi (\gamma, \delta, \varepsilon, \dots \lambda) \dots (6).$$

But from (5) we have

$$\phi (\nu, \mu, \lambda, \dots \varepsilon, \delta, \gamma) = \nu.\phi (\mu, \lambda, \dots \varepsilon, \delta, \gamma) + \phi (\lambda, \dots \varepsilon, \delta, \gamma) \dots (7).$$

Hence, it is clear by a comparison of (6) and (7) that if we can prove that

$$\phi (\gamma, \delta, \varepsilon, \dots \lambda, \mu) = \phi (\mu, \lambda, \dots \varepsilon, \delta, \gamma),$$

and that $\quad \phi (\gamma, \delta, \varepsilon, \dots \lambda) \quad = \quad \phi (\lambda, \dots \varepsilon, \delta, \gamma),$

it will follow at once that the same property holds if we consider an additional quantity ν; that is to say, we shall thus prove that

$$\phi (\gamma, \delta, \varepsilon, \dots \lambda, \mu, \nu) = \phi (\nu, \mu, \lambda, \dots \varepsilon, \delta, \gamma).$$

But we have already shown that

$$\phi (\gamma, \delta) = \gamma \delta + 1 = \phi (\delta, \gamma),$$

and that $\quad \phi (\gamma, \delta, \varepsilon) = \varepsilon \delta \gamma + \varepsilon + \gamma = \phi (\varepsilon, \delta, \gamma).$

Hence, it follows by induction that the general theorem is true; namely, that

$$\phi (\gamma, \delta, \varepsilon, \dots \lambda, \mu, \nu) = \phi (\nu, \mu, \lambda, \dots \varepsilon, \delta, \gamma),$$

however many there may be of the quantities

$$\gamma, \delta, \varepsilon, \dots \lambda, \mu, \nu.$$

It will be seen that the relations (1) are those which hold between the denominators of successive convergents to a continued fraction, and that the theorem here proved is the first of those quoted in Art. 94.

3. We have also

$$\frac{d}{d\nu}\,\phi\,(\gamma,\,\delta,\,\varepsilon,\,\dots\,\lambda,\,\mu,\,\nu) = \frac{d}{d\nu}\,\phi\,(\nu,\,\mu,\,\lambda,\,\dots\,\varepsilon,\,\delta,\,\gamma)$$

$$= \frac{d}{d\nu}\,\{\nu.\phi(\mu,\,\lambda,\dots\varepsilon,\delta,\gamma)+\phi(\lambda,\dots\varepsilon,\delta,\gamma)\}$$

$$= \phi\,(\mu,\,\lambda,\,\dots\,\varepsilon,\,\delta,\,\gamma)$$

$$= \phi\,(\gamma,\,\delta,\,\varepsilon,\,\dots\,\lambda,\,\mu)\dots\dots\dots\dots(1).$$

In a similar manner we get

$$\frac{d}{d\mu}\,\phi\,(\gamma,\,\delta,\,\varepsilon,\,\dots\,\lambda,\,\mu) = \frac{d}{d\mu}\,\phi\,(\mu,\,\lambda,\,\dots\,\varepsilon,\,\delta,\,\gamma)$$

$$= \phi\,(\lambda,\,\dots\,\varepsilon,\,\delta,\,\gamma)$$

$$= \phi\,(\gamma,\,\delta,\,\varepsilon,\,\dots\,\lambda).$$

Therefore

$$\frac{d^{u}}{d\mu\,d\nu}\,\phi\,(\gamma,\,\delta,\,\varepsilon,\,\dots\,\lambda,\,\mu,\,\nu) = \phi\,(\gamma,\,\delta,\,\varepsilon,\,\dots\,\lambda)\dots.(2).$$

These are the other theorems quoted in Art. 94.

4. In the two preceding articles the theorems have been considered in a general form, but they may be proved very readily by considering the value of A expressed as a determinant.

We have for n surfaces

$$A = - \begin{vmatrix} -\rho_1, & -1, & 0, & 0, & 0, & \dots\dots\dots \\ 1, & -t_1, & -1, & 0, & 0, & \dots\dots\dots \\ 0, & 1, & -\rho_2, & -1, & 0, & \dots\dots\dots \\ 0, & 0, & 1, & -t_2, & -1, & \dots\dots\dots \\ \multicolumn{6}{c}{\dots\dots\dots\dots\dots\dots\dots\dots\dots\dots\dots} \\ \multicolumn{6}{c}{\dots\dots\dots\dots\dots\dots\dots, \; 1, \; -t_{n-1}, \; -1} \\ \multicolumn{6}{c}{\dots\dots\dots\dots\dots\dots\dots\dots \; 1 \;, \; -\rho_n} \end{vmatrix}.$$

And we may show by an even number of transferences of columns and lines that the above determinant

$$= \begin{vmatrix} -\rho_n, & -1 & , & 0 & , & 0 & , & 0, \ldots\ldots\ldots \\ 1, & -t_{n-1}, & -1 & , & 0 & , & 0, \ldots\ldots\ldots \\ 0, & 1 & , & -\rho_{n-1}, & -1 & , & 0, \ldots\ldots\ldots \\ 0, & 0 & , & 1 & , & -t_{n-1}, & -1, \ldots\ldots\ldots \\ \ldots\ldots\ldots\ldots\ldots\ldots\ldots\ldots\ldots\ldots\ldots\ldots\ldots\ldots \\ \ldots\ldots\ldots\ldots\ldots\ldots\ldots\ldots\ldots, & 1, & -t_1, & -1 \\ \ldots\ldots\ldots\ldots\ldots\ldots\ldots\ldots\ldots & 1, & -\rho_1 \end{vmatrix};$$

an equality which proves Theorem I.

5. Again we know by definition and analogy that

$$B = - \begin{vmatrix} -t_1, & -1, & 0, & 0, & \ldots\ldots\ldots\ldots \\ 1, & -\rho_2, & -1, & 0, & \ldots\ldots\ldots\ldots \\ 0, & 1, & -t_2, & -1, & \ldots\ldots\ldots\ldots \\ \ldots\ldots\ldots\ldots\ldots\ldots\ldots\ldots\ldots\ldots\ldots \\ \ldots\ldots\ldots\ldots\ldots, & 1, & -t_{n-1}, & -1 \\ \ldots\ldots\ldots\ldots\ldots & 1, & -\rho_n \end{vmatrix}$$

$$= \frac{dA}{d\rho_1}.$$

Similarly $$C = \frac{dA}{d\rho_n},$$

and $$D = \frac{d^2 A}{d\rho_1 d\rho_n}.$$

PRINTED BY W. METCALFE AND SON, TRINITY STREET, CAMBRIDGE.

A CLASSIFIED LIST

OF

EDUCATIONAL WORKS

PUBLISHED BY

GEORGE BELL & SONS.

Full Catalogues will be sent post free on application.

BIBLIOTHECA CLASSICA.

A Series of Greek and Latin Authors, with English Notes, edited by eminent Scholars. 8vo.

Æschylus. By F. A. Paley, M.A. 18*s.*

Cicero's Orations. By G. Long, M.A. 4 vols. 16*s.*, 14*s.*, 16*s.*, 18*s.*

Demosthenes. By R. Whiston, M.A. 2 vols. 16*s.* each.

Euripides. By F. A. Paley, M.A. 3 vols. 16*s.* each.

Homer. By F. A. Paley, M.A. Vol. I. 12*s.*; Vol. II. 14*s.*

Herodotus. By Rev. J. W. Blakesley, B.D. 2 vols. 32*s.*

Hesiod. By F. A. Paley, M.A. 10*s.* 6*d.*

Horace. By Rev. A. J. Macleane, M.A. 18*s.*

Juvenal and Persius. By Rev. A. J. Macleane, M.A. 12*s.*

Lucan. The Pharsalia. By C. E. Haskins, M.A. [*In the press.*

Plato. By W. H. Thompson, D.D. 2 vols. 7*s.* 6*d.* each.

Sophocles. Vol. I. By Rev. F. H. Blaydes, M.A. 18*s.*

———— Vol. II. Philoctetes. Electra. Ajax and Trachiniæ. By F. A. Paley, M.A. 12*s.*

Tacitus: The Annals. By the Rev. P. Frost. 15*s.*

Terence. By E. St. J. Parry, M.A. 18*s.*

Virgil. By J. Conington, M.A. 3 vols. 14*s.* each.

An Atlas of Classical Geography; Twenty-four Maps. By W. Hughes and George Long, M.A. New edition, with coloured Outlines. Imperial 8vo. 12*s.* 6*d.*

Uniform with above.

A Complete Latin Grammar. By J. W. Donaldson, D.D. 3rd Edition. 14*s.*

GRAMMAR-SCHOOL CLASSICS.

A Series of Greek and Latin Authors, with English Notes. Fcap. 8vo.

Cæsar: De Bello Gallico. By George Long, M.A. 5*s.* 6*d.*

———— Books I.–III. For Junior Classes. By G. Long, M.A. 2*s.* 6*d.*

———— Books IV. and V. in 1 vol. 1*s.* 6*d.*

Catullus, Tibullus, and Propertius. Selected Poems. With Life. By Rev. A. H. Wratislaw. 3*s.* 6*d.*

Cicero: De Senectute, De Amicitia, and Select Epistles. By George Long, M.A. 4s. 6d.

Cornelius Nepos. By Rev. J. F. Macmichael. 2s. 6d.

Homer: Iliad. Books I.-XII. By F. A. Paley, M.A. 6s. 6d.

Horace. With Life. By A. J. Macleane, M.A. 6s. 6d. [In 2 parts, 3s. 6d. each.]

Juvenal: Sixteen Satires. By H. Prior, M.A. 4s. 6d.

Martial: Select Epigrams. With Life. By F. A. Paley, M.A. 6s. 6d.

Ovid: the Fasti. By F. A. Paley, M.A. 5s.

Sallust: Catilina and Jugurtha. With Life. By G. Long, M.A. and J. G. Frazer. 5s., or separately, 2s. 6d. each.

Tacitus: Germania and Agricola. By Rev. P. Frost. 3s. 6d.

Virgil: Bucolics, Georgics, and Æneid, Books I.-IV. Abridged from Professor Conington's Edition. 5s. 6d.—Æneid, Books V.-XII. 5s. 6d. Also in 9 separate Volumes, 1s. 6d. each.

Xenophon: The Anabasis. With Life. By Rev. J. F. Macmichael. 5s. Also in 4 separate volumes, 1s. 6d. each.

———— The Cyropædia. By G. M. Gorham, M.A. 6s.

———— Memorabilia. By Percival Frost, M.A. 4s. 6d.

A Grammar-School Atlas of Classical Geography, containing Ten selected Maps. Imperial 8vo. 5s.

Uniform with the Series.

The New Testament, in Greek. With English Notes, &c. By Rev. J. F. Macmichael. 7s. 6d.

CAMBRIDGE GREEK AND LATIN TEXTS.

Æschylus. By F. A. Paley, M.A. 3s.

Cæsar: De Bello Gallico. By G. Long, M.A. 2s.

Cicero: De Senectute et de Amicitia, et Epistolæ Selectæ. By G. Long, M.A. 1s. 6d.

Ciceronis Orationes. Vol. I. (in Verrem.) By G. Long, M.A. 3s. 6d.

Euripides. By F. A. Paley, M.A. 3 vols. 3s. 6d. each.

Herodotus. By J. G. Blakesley, B.D. 2 vols. 7s.

Homeri Ilias. I.-XII. By F. A. Paley, M.A. 2s. 6d.

Horatius. By A. J. Macleane, M.A. 2s. 6d.

Juvenal et Persius. By A. J. Macleane, M.A. 1s. 6d.

Lucretius. By H. A. J. Munro, M.A. 2s. 6d.

Sallusti Crispi Catilina et Jugurtha. By G. Long, M.A. 1s. 6d.

Sophocles. By F. A. Paley, M.A. 3s. 6d.

Terenti Comœdiæ. By W. Wagner, Ph.D. 3s.

Thucydides. By J. G. Donaldson, D.D. 2 vols. 7s.

Virgilius. By J. Conington, M.A. 3s. 6d.

Xenophontis Expeditio Cyri. By J. F. Macmichael, B.A. 2s. 6d.

Novum Testamentum Græcum. By F. H. Scrivener, M.A. 4s. 6d. An edition with wide margin for notes, half bound, 12s.

CAMBRIDGE TEXTS WITH NOTES.

A Selection of the most usually read of the Greek and Latin Authors, Annotated for Schools. Fcap. 8vo. 1s. 6d. each, with exceptions.

Euripides. Alcestis.—Medea.—Hippolytus.—Hecuba.—Bacchæ.
Ion. 2s. —Orestes. — Phoenissæ.—Troades. — Hercules Furens.—Andromache.—Iphigenia in Tauris. By F. A. Paley, M.A.

Æschylus. Prometheus Vinctus.—Septem contra Thebas.—Agamemnon.—Persæ.—Eumenides. By F. A. Paley, M.A.

Sophocles. Œdipus Tyrannus.—Œdipus Coloneus.—Antigone.
By F. A. Paley, M.A.

Homer. Iliad. Book I. By F. A. Paley, M.A. 1s.

Terence. Andria.—Hauton Timorumenos.—Phormio.—Adelphoe.
By Professor Wagner, Ph.D.

Cicero. De Senectute, De Amicitia, and Epistolæ Selectæ. By G. Long, M.A.

Ovid. Selections. By A. J. Macleane, M.A.
Others in preparation.

PUBLIC SCHOOL SERIES.

A Series of Classical Texts, annotated by well-known Scholars. Cr. 8vo.

Aristophanes. The Peace. By F. A. Paley, M.A. 4s. 6d.
———— The Acharnians. By F. A. Paley, M.A. 4s. 6d.
———— The Frogs. By F. A. Paley, M.A. 4s. 6d.

Cicero. The Letters to Atticus. Bk. I. By A. Pretor, M.A. 4s. 6d.

Demosthenes de Falsa Legatione. By R. Shilleto, M.A. 6s.
———— The Law of Leptines. By B. W. Beatson, M.A. 3s. 6d.

Livy. Book XXI. Edited, with Introduction, Notes, and Maps, by the Rev. L. D. Dowdall, M.A., B.D. 3s. 6d.

Plato. The Apology of Socrates and Crito. By W. Wagner, Ph.D. 8th Edition. 4s. 6d.
———— The Phædo. 7th Edition. By W. Wagner, Ph.D. 5s. 6d.
———— The Protagoras. 4th Edition. By W. Wayte, M.A. 4s. 6d.
———— The Euthyphro. 2nd Edition. By G. H. Wells, M.A. 3s.
———— The Euthydemus. By G. H. Wells, M.A. 4s.
———— The Republic. Books I. & II. By G. H. Wells, M.A. 5s. 6d.

Plautus. The Aulularia. By W. Wagner, Ph.D. 3rd Edition. 4s. 6d.
———— Trinummus. By W. Wagner, Ph.D. 3rd Edition. 4s. 6d.
———— The Menaechmei. By W. Wagner, Ph.D. 4s. 6d.
———— The Mostellaria. By Prof. E. A. Sonnenschein. 5s.

Sophoclis Trachiniæ. By A. Pretor, M.A. 4s. 6d.

Sophocles. Oedipus Tyrannus. By B. H. Kennedy, D.D. 5s.

Terence. By W. Wagner, Ph.D. 10s. 6d.

Theocritus. By F. A. Paley, M.A. 4s. 6d.

Thucydides. Book VI. By T. W. Dougan, M.A., Fellow of St. John's College, Cambridge. 6s.
Others in preparation.

CRITICAL AND ANNOTATED EDITIONS.

Ætna. By H. A. J. Munro, M.A. 3s. 6d.

Aristophanis Comœdiæ. By H. A. Holden, LL.D. 8vo. 2 vols. 23s. 6d. Plays sold separately.
———— Pax. By F. A. Paley, M.A. Fcap. 8vo. 4s. 6d.

Corpus Poetarum Latinorum. Edited by Walker. 1 vol. 8vo. 18s.

Horace. Quinti Horatii Flacci Opera. By H. A. J. Munro, M.A.
Large 8vo. 1l. 1s.
Livy. The first five Books. By J. Prendeville. 12mo. roan, 5s.
Or Books I.-III. 3s. 6d. IV. and V. 3s. 6d.
Lucretius. With Commentary by H. A. J. Munro. Immediately.
New Edition.
Ovid. P. Ovidii Nasonis Heroides XIV. By A. Palmer, M.A. 8vo. 6s.
———— P. Ovidii Nasonis Ars Amatoria et Amores. By the Rev.
Herb. Williams, M.A. 3s. 6d. [2s. 6d.
———— Metamorphoses. Book XIII. By Chas. Haines Keane, M.A.
Propertius. Sex Aurelii Propertii Carmina. By F. A. Paley, M.A.
8vo. Cloth, 9s.
Sex Propertii Elegiarum. Libri IV. By A. Palmer. Fcap. 8vo. 5s.
Sophocles. The Ajax. By C. E. Palmer, M.A. 4s. 6d.
———— The Oedipus Tyrannus. By B. H. Kennedy, D.D. With
a Commentary, containing selected Notes by the late T. H. Steel, M.A.
Crown 8vo. 8s.
Thucydides. The History of the Peloponnesian War. By Richard
Shilleto, M.A. Book I. 8vo. 6s. 6d. Book II. 8vo. 5s. 6d.

LATIN AND GREEK CLASS-BOOKS.

Auxilia Latina. A Series of Progressive Latin Exercises. By
M. J. B. Baddeley, M.A. Fcap. 8vo. Part I. Accidence. 2nd Edition, revised.
2s. Part II. 4th Edition, revised. 2s. Key to Part II. 2s. 6d.
Latin Prose Lessons. By Prof. Church, M.A. 6th Edit. Fcap. 8vo.
2s. 6d.
Latin Exercises and Grammar Papers. By T. Collins, M.A. 5th
Edition. Fcap. 8vo. 2s. 6d.
Unseen Papers in Latin Prose and Verse. With Examination
Questions. By T. Collins, M.A. 3rd Edition. Fcap. 8vo. 2s. 6d.
———— in Greek Prose and Verse. With Examination Questions.
By T. Collins, M.A. 2nd Edition. Fcap. 8vo. 3s.
Tales for Latin Prose Composition. With Notes and Vocabu-
lary. By G. H. Wells, M.A. 2s.
Latin Vocabularies for Repetition. By A. M. M. Stedman, M.A.
Fcap. 8vo. 1s. 6d.
Analytical Latin Exercises. By C. P. Mason, B.A. 4th Edit.
Part I., 1s. 6d. Part II., 2s. 6d.
Latin Mood Construction, Outlines of. With Exercises. By
the Rev. G. E. C. Casey, M.A., F.L.S., F.G.S. Small post 8vo. 1s. 6d.
Latin of the Exercises. 1s. 6d.
Scala Latina. Elementary Latin Exercises. By Rev. J. W.
Davis, M.A. New Edition, with Vocabulary. Fcap. 8vo. 2s. 6d.
Scala Græca: a Series of Elementary Greek Exercises. By Rev. J. W.
Davis, M.A., and R. W. Baddeley, M.A. 3rd Edition. Fcap. 8vo. 2s. 6d.
Greek Verse Composition. By G. Preston, M.A. Crown 8vo. 4s. 6d.
Greek Particles and their Combinations according to Attic Usage.
A Short Treatise. By F. A. Paley, M.A. 2s. 6d.

BY THE REV. P. FROST, M.A., ST. JOHN'S COLLEGE, CAMBRIDGE.

Eclogæ Latinæ; or, First Latin Reading-Book, with English Notes
and a Dictionary. New Edition. Fcap. 8vo. 2s. 6d.
Materials for Latin Prose Composition. New Edition. Fcap. 8vo.
2s. 6d. Key, 4s.
A Latin Verse-Book. An Introductory Work on Hexameters and
Pentameters. New Edition. Fcap. 8vo. 3s. Key, 5s.
Analecta Græca Minora, with Introductory Sentences, English
Notes, and a Dictionary. New Edition. Fcap. 8vo. 3s. 6d.

Materials for Greek Prose Composition. New Edit. Fcap. 8vo.
3s. 6d. Key, 5s.
Florilegium Poeticum. Elegiac Extracts from Ovid and Tibullus.
New Edition. With Notes. Fcap. 8vo. 3s.
Anthologia Græca. A Selection of Choice Greek Poetry, with Notes.
By F. St. John Thackeray. *4th and Cheaper Edition.* 16mo. 4s. 6d.
Anthologia Latina. A Selection of Choice Latin Poetry, from
Nævius to Boëthius, with Notes. By Rev. F. St. John Thackeray. Revised
and Cheaper Edition. 16mo. 4s. 6d.

BY H. A. HOLDEN, LL.D.

Foliorum Silvula. Part I. Passages for Translation into Latin
Elegiac and Heroic Verse. 10th Edition. Post 8vo. 7s. 6d.
———— Part II. Select Passages for Translation into Latin Lyric
and Comic Iambic Verse. 3rd Edition. Post 8vo. 5s.
———— Part III. Select Passages for Translation into Greek Verse.
3rd Edition. Post 8vo. 8s.
**Folia Silvulæ, sive Eclogæ Poetarum Anglicorum in Latinum et
Græcum conversæ.** 8vo. Vol. II. 12s.
Foliorum Centuriæ. Select Passages for Translation into Latin
and Greek Prose. 9th Edition. Post 8vo. 8s.

TRANSLATIONS, SELECTIONS, &c.

⁎ Many of the following books are well adapted for School Prizes.
Æschylus. Translated into English Prose by F. A. Paley, M.A.
2nd Edition. 8vo. 7s. 6d.
———— Translated into English Verse by Anna Swanwick. Post
8vo. 5s.
Homer. The Iliad. Books I.-IV. Translated into English
Hexameter Verse by Henry Smith Wright, B.A. Royal 8vo. 5s.
Horace. The Odes and Carmen Sæculare. In English Verse by
J. Conington, M.A. 9th edition. Fcap. 8vo. 5s. 6d.
———— The Satires and Epistles. In English Verse by J. Coning-
ton, M.A. 6th edition. 6s. 6d.
———— Illustrated from Antique Gems by C. W. King, M.A. The
text revised with Introduction by H. A. J. Munro, M.A. Large 8vo. 1l. 1s.
———— Translations from. By Sir Stephen E. de Vere, Bart.,
with Latin Text. Crown 8vo. 3s. 6d.
Horace's Odes. Englished and Imitated by various hands. Edited
by C. W. F. Cooper. Crown 8vo. 6s. 6d.
Lusus Intercisi. Verses, Translated and Original, by H. J.
Hodgson, M.A., formerly Fellow of Trinity College, Cambridge. 5s.
Propertius. Verse Translations from Book V., with revised Latin
Text. By F. A. Paley, M.A. Fcap. 8vo. 3s.
Plato. Gorgias. Translated by E. M. Cope, M.A. 8vo. 7s.
———— Philebus. Translated by F. A. Paley, M.A. Small 8vo. 4s.
———— Theætetus. Translated by F. A. Paley, M.A. Small 8vo. 4s.
———— Analysis and Index of the Dialogues. By Dr. Day. Post 8vo. 5s.
Reddenda Reddita : Passages from English Poetry, with a Latin
Verse Translation. By F. E. Gretton. Crown 8vo. 6s.
Sabrinæ Corolla in Hortulis Regiæ Scholæ Salopiensis contexuerunt
tres viri floribus legendis. Editio tertia. 8vo. 8s. 6d.
Theocritus. In English Verse, by C. S. Calverley, M.A. New
Edition, revised. Crown 8vo. 7s. 6d.
Translations into English and Latin. By C. S. Calverley, M.A.
Post 8vo. 7s. 6d.

Translations into Greek and Latin Verse. By R. C. Jebb. 4to. cloth gilt. 10s. 6d.
———— into English, Latin, and Greek. By R. C. Jebb, M.A., H. Jackson, Litt.D., and W. E. Currey, M.A. Second Edition. 8s.
Between Whiles. Translations by Rev. B. H. Kennedy, D.D. 2nd Edition, revised. Crown 8vo. 5s.

REFERENCE VOLUMES.

A Latin Grammar. By Albert Harkness. Post 8vo. 6s.
———— By T. H. Key, M.A. 6th Thousand. Post 8vo. 8s.
A Short Latin Grammar for Schools. By T. H. Key, M.A. F.R.S. 15th Edition. Post 8vo. 3s. 6d.
A Guide to the Choice of Classical Books. By J. B. Mayor, M.A. 3rd Edition, with a Supplementary List. Crown 8vo. 4s. 6d. Supplementary List, 1s. 6d.
The Theatre of the Greeks. By J. W. Donaldson, D.D. 8th Edition. Post 8vo. 5s.
Keightley's Mythology of Greece and Italy. 4th Edition. 5s.
A Dictionary of Latin and Greek Quotations. By H. T. Riley. Post 8vo. 5s. With Index Verborum, 6s.
A History of Roman Literature. By W. S. Teuffel, Professor at the University of Tübingen. By W. Wagner, Ph.D. 2 vols. Demy 8vo. 21s.
Student's Guide to the University of Cambridge. 4th Edition revised. Fcap. 8vo. 6s. 6d.; or in Parts.—Part 1, 2s. 6d.; Parts 2 to 9, 1s. each.

CLASSICAL TABLES.

Latin Accidence. By the Rev. P. Frost, M.A. 1s.
Latin Versification. 1s.
Notabilia Quædam; or the Principal Tenses of most of the Irregular Greek Verbs and Elementary Greek, Latin, and French Construction. New Edition. 1s.
Richmond Rules for the Ovidian Distich, &c. By J. Tate, M.A. 1s.
The Principles of Latin Syntax. 1s.
Greek Verbs. A Catalogue of Verbs, Irregular and Defective; their leading formations, tenses, and inflexions, with Paradigms for conjugation, Rules for formation of tenses, &c. &c. By J. S. Baird, T.C.D. 2s. 6d.
Greek Accents (Notes on). By A. Barry, D.D. New Edition. 1s.
Homeric Dialect. Its Leading Forms and Peculiarities. By J. S. Baird, T.C.D. New Edition, by W. G. Rutherford. 1s.
Greek Accidence. By the Rev. P. Frost, M.A. New Edition. 1s.

CAMBRIDGE MATHEMATICAL SERIES.

Algebra. Choice and Chance. By W. A. Whitworth, M.A. 3r Edition. 6s.
Euclid. Exercises on Euclid and in Modern Geometry. By J. McDowell, M.A. 3rd Edition. 6s.
Trigonometry. Plane. By Rev. T. Vyvyan, M.A. 3rd Edit. 3s. 6d.
Geometrical Conic Sections. By H. G. Willis, M.A. Manchester Grammar School. 7s. 6d.
Conics. The Elementary Geometry of. 4th Edition. By C. Taylor, D.D. 4s. 6d.

Solid Geometry. By W. S. Aldis, M.A. 3rd Edition. 6*s*.
Rigid Dynamics. By W. S. Aldis, M.A. 4*s*.
Elementary Dynamics. By W. Garnett, M.A. 3rd Edition. 6*s*.
Dynamics. A Treatise on. By W. H. Besant, D.Sc., F.R.S. 7*s*. 6*d*.
Heat. An Elementary Treatise. By W. Garnett, M.A. 3rd Edit.
3*s*. 6*d*.
Hydromechanics. By W. H. Besant, M.A., F.R.S. 4th Edition.
Part I. Hydrostatics. 5*s*.
Mechanics. Problems in Elementary. By W. Walton, M.A. 6*s*.

CAMBRIDGE SCHOOL AND COLLEGE TEXT-BOOKS.

A Series of Elementary Treatises for the use of Students in the
Universities, Schools, and Candidates for the Public
Examinations. Fcap. 8vo.

Arithmetic. By Rev. C. Elsee, M.A. Fcap. 8vo. 12th Edit. 3*s*. 6*d*.
Algebra. By the Rev. C. Elsee, M.A. 6th Edit. 4*s*.
Arithmetic. By A. Wrigley, M.A. 3*s*. 6*d*.
———— A Progressive Course of Examples. With Answers. By
J. Watson, M.A. 5th Edition. 2*s*. 6*d*.
Algebra. Progressive Course of Examples. By Rev. W. F.
M'Michael, M.A., and R. Prowde Smith, M.A. 4th Edition. 3*s*.6*d*. With
Answers. 4*s*. 6*d*.
Plane Astronomy, An Introduction to. By P. T. Main, M.A.
5th Edition. 4*s*.
Conic Sections treated Geometrically. By W. H. Besant, M.A.
5th Edition. 4*s*. 6*d*. Solution to the Examples. 4*s*.
Elementary Conic Sections treated Geometrically. By W. H.
Besant, M.A. [*In the press.*
Conics. Enunciations and Figures. By W. H. Besant, M.A. 1*s*. 6*d*.
Statics, Elementary. By Rev. H. Goodwin, D.D. 2nd Edit. *s*.
Hydrostatics, Elementary. By W. H. Besant, M.A. 11th Edit 4*s*.
Mensuration, An Elementary Treatise on. By B.T. Moore, M.A 6*s*.
Newton's Principia, The First Three Sections of, with an Appen-
dix; and the Ninth and Eleventh Sections. By J. H. Evans, M.A. 5th
Edition, by P. T. Main, M.A. 4*s*.
Optics, Geometrical. With Answers. By W. S. Aldis, M.A. 3*s*. 6*d*.
Analytical Geometry for Schools. By T. G. Vyvyan. 4th Edit. 4*s*. 6*d*.
Greek Testament, Companion to the. By A. C. Barrett, A.M.
5th Edition, revised. Fcap. 8vo. 5*s*.
Book of Common Prayer, An Historical and Explanatory Treatise
on the. By W. G. Humphry, B.D. 6th Edition. Fcap. 8vo. 2*s*. 6*d*.
Music, Text-book of. By H. C. Banister. 12th Edit. revised. 5*s*.
———— Concise History of. By Rev. H. G. Bonavia Hunt, B. Mus.
Oxon. 7th Edition revised. 3*s*. 6*d*.

ARITHMETIC AND ALGEBRA.
See foregoing Series.

GEOMETRY AND EUCLID.

Euclid. The Definitions of, with Explanations and Exercises, and an Appendix of Exercises on the First Book. By R. Webb, M.A. Crown 8vo. 1s. 6d.

────── Book I. With Notes and Exercises for the use of Preparatory Schools, &c. By Braithwaite Arnett, M.A. 8vo. 4s. 6d.

────── The First Two Books explained to Beginners. By C. P. Mason, B.A. 2nd Edition. Fcap. 8vo. 2s. 6d.

The Enunciations and Figures to Euclid's Elements. By Rev. J. Brasse, D.D. New Edition. Fcap. 8vo. 1s. On Cards, in case, 5s. Without the Figures, 6d.

Exercises on Euclid and in Modern Geometry. By J. McDowell, B.A. Crown 8vo. 3rd Edition revised. 6s.

Geometrical Conic Sections. By H. G. Willis, M.A. 7s. 6d.

Geometrical Conic Sections. By W. H. Besant, M.A. 5th Edit. 4s. 6d. Solution to the Examples. 4s.

Elementary Geometrical Conic Sections. By W. H. Besant, M.A. [In the press.]

Elementary Geometry of Conics. By C. Taylor, D.D. 4th Edit. 8vo. 4s. 6d.

An Introduction to Ancient and Modern Geometry of Conics. By C. Taylor, M.A. 8vo. 15s.

Solutions of Geometrical Problems, proposed at St. John's College from 1830 to 1846. By T. Gaskin, M.A. 8vo. 12s.

TRIGONOMETRY.

Trigonometry, Introduction to Plane. By Rev. T. G. Vyvyan, Charterhouse. 3rd Edition. Cr. 8vo. 3s. 6d.

An Elementary Treatise on Mensuration. By B. T. Moore, M.A. 5s.

ANALYTICAL GEOMETRY AND DIFFERENTIAL CALCULUS.

An Introduction to Analytical Plane Geometry. By W. P. Turnbull, M.A. 8vo. 12s.

Problems on the Principles of Plane Co-ordinate Geometry. By W. Walton, M.A. 8vo. 16s.

Trilinear Co-ordinates, and Modern Analytical Geometry of Two Dimensions. By W. A. Whitworth, M.A. 8vo. 16s.

An Elementary Treatise on Solid Geometry. By W. S. Aldis, M.A. 3rd Edition revised. Cr. 8vo. 6s.

Elementary Treatise on the Differential Calculus. By M. O'Brien, M.A. 8vo. 10s. 6d.

Elliptic Functions, Elementary Treatise on. By A. Cayley, M.A. Demy 8vo. 15s.

MECHANICS & NATURAL PHILOSOPHY.

Statics, Elementary. By H. Goodwin, D.D. Fcap. 8vo. 2nd Edition. 3s.

Dynamics, A Treatise on Elementary. By W. Garnett, M.A. 3rd Edition. Crown 8vo. 6s.

Dynamics. Rigid. By W. S. Aldis, M.A. 4*s.*

Dynamics. A Treatise on. By W. H. Besant, D.Sc., F.R.S. 7*s.* 6*d.*

Elementary Mechanics, Problems in. By W. Walton, M.A. New Edition. Crown 8vo. 6*s.*

Theoretical Mechanics, Problems in. By W. Walton. 2nd Edit. revised and enlarged. Demy 8vo. 16*s.*

Hydrostatics. By W. H. Besant, M.A. Fcap. 8vo. 11th Edition. 4*s.*

Hydromechanics, A Treatise on. By W. H. Besant, M.A., F.R.S. 8vo. 4th Edition, revised. Part I. Hydrostatics. 5*s.*

Optics, Geometrical. By W. S. Aldis, M.A. Fcap. 8vo. 3*s.* 6*d.*

Double Refraction, A Chapter on Fresnel's Theory of. By W. S. Aldis, M.A. 8vo. 2*s.*

Heat, An Elementary Treatise on. By W. Garnett, M.A. Crown 8vo. 3rd Edition revised. 3*s.* 6*d.*

Newton's Principia, The First Three Sections of, with an Appendix; and the Ninth and Eleventh Sections. By J. H. Evans, M.A. 5th Edition. Edited by P. T. Main, M.A. 4*s.*

Astronomy, An Introduction to Plane. By P. T. Main, M.A. Fcap. 8vo. cloth. 4*s.*

Astronomy, Practical and Spherical. By R. Main, M.A. 8vo. 14*s.*

Astronomy, Elementary Chapters on, from the 'Astronomie Physique' of Biot. By H. Goodwin, D.D. 8vo. 3*s.* 6*d.*

Pure Mathematics and Natural Philosophy, A Compendium of Facts and Formulæ in. By G. R. Smalley. 2nd Edition, revised by J. McDowell, M.A. Fcap. 8vo. 3*s.* 6*d.*

Elementary Mathematical Formulæ. By the Rev. T. W. Openshaw. 1*s.* 6*d.*

Elementary Course of Mathematics. By H. Goodwin, D.D. 6th Edition. 8vo. 16*s.*

Problems and Examples, adapted to the 'Elementary Course of Mathematics.' 3rd Edition. 8vo. 5*s.*

Solutions of Goodwin's Collection of Problems and Examples. By W. W. Hutt, M.A. 3rd Edition, revised and enlarged. 8vo. 9*s.*

Mechanics of Construction. With numerous Examples. By S. Fenwick, F.R.A.S. 8vo. 12*s.*

TECHNOLOGICAL HANDBOOKS.
Edited by H. Trueman Wood, Secretary of the Society of Arts.

1. Dyeing and Tissue Printing. By W. Crookes, F.R.S. 5*s.*

2. Glass Manufacture. By Henry Chance, M.A.; H. J. Powell, B.A.; and H. G. Harris. 3*s.* 6*d.*

3. Cotton Manufacture. By Richard Marsden, Esq., of Manchester. 6*s.* 6*d.*

4. Chemistry of Coal-Tar Colours. By Prof. Benedikt. Translated by Dr. Knecht of Bradford. 5*s.*

HISTORY, TOPOGRAPHY, &c.

Rome and the Campagna. By R. Burn, M.A. With 85 Engravings and 26 Maps and Plans. With Appendix. 4to. 3*l.* 3*s.*

Old Rome. A Handbook for Travellers. By R. Burn, M.A. With Maps and Plans. Demy 8vo. 10*s.* 6*d.*

Modern Europe. By Dr. T. H. Dyer. 2nd Edition, revised and continued. 5 vols. Demy 8vo. 2*l.* 12*s.* 6*d.*

The History of the Kings of Rome. By Dr. T. H. Dyer. 8vo. 16*s.*

The History of Pompeii: its Buildings and Antiquities. By T. H. Dyer. 3rd Edition, brought down to 1874. Post 8vo. 7*s.* 6*d.*

The City of Rome: its History and Monuments. 2nd Edition revised by T. H. Dyer. 5*s.*

Ancient Athens: its History, Topography, and Remains. By T. H. Dyer. Super-royal 8vo. Cloth. 1*l.* 5*s.*

The Decline of the Roman Republic. By G. Long. 5 vols. 8vo. 14*s.* each.

A History of England during the Early and Middle Ages. By C. H. Pearson, M.A. 2nd Edition revised and enlarged. 8vo. Vol. I. 16*s.* Vol. II. 14*s.*

Historical Maps of England. By C. H. Pearson. Folio. 3rd Edition revised. 31*s.* 6*d.*

History of England, 1800–15. By Harriet Martineau, with new and copious Index. 1 vol. 3*s.* 6*d.*

History of the Thirty Years' Peace, 1815–46. By Harriet Martineau. 4 vols. 3*s.* 6*d.* each.

A Practical Synopsis of English History. By A. Bowes. 4th Edition. 8vo. 2*s.*

Lives of the Queens of England. By A. Strickland. Library Edition, 8 vols. 7*s.* 6*d.* each. Cheaper Edition, 6 vols. 5*s.* each. Abridged Edition, 1 vol. 6*s.* 6*d.*

Eginhard's Life of Karl the Great (Charlemagne). Transla with Notes, by W. Glaister, M.A., B.C.L. Crown 8vo. 4*s.* 6*d.*

Outlines of Indian History. By A. W. Hughes. Small Po 8vo. 3*s.* 6*d.*

The Elements of General History. By Prof. Tytler. New Edition, brought down to 1874. Small Post 8vo. 3*s.* 6*d.*

ATLASES.

An Atlas of Classical Geography. 24 Maps. By W. Hugh and G. Long, M.A. New Edition. Imperial 8vo. 12*s.* 6*d.*

A Grammar-School Atlas of Classical Geography. Ten Map selected from the above. New Edition. Imperial 8vo. 5*s.*

First Classical Maps. By the Rev. J. Tate, M.A. 3rd Edition. Imperial 8vo. 7*s.* 6*d.*

Standard Library Atlas of Classical Geography. Imp. 8vo. 7*s.* 6*d.*

PHILOLOGY.

WEBSTER'S DICTIONARY OF THE ENGLISH LANGUAGE. With Dr. Mahn's Etymology. 1 vol. 1628 pages, 3000 Illustrations. 21*s.* With Appendices and 70 additional pages of Illustrations, 1919 pages, 31*s.* 6*d.*

'THE BEST PRACTICAL ENGLISH DICTIONARY EXTANT.'—*Quarterly Review*, 1873.

Prospectuses, with specimen pages, post free on application.

Richardson's Philological Dictionary of the English Language. Combining Explanation with Etymology, and copiously illustrated by Quotations from the best Authorities. With a Supplement. 2 vols. 4to. 4*l.* 14*s.* 6*d.*; half russia, 5*l.* 15*s.* 6*d.*; russia, 6*l.* 12*s.* Supplement separately. 4to. 12*s.*

An 8vo. Edit. without the Quotations, 15*s.*; half russia, 20*s.*; russia, 24*s.*

Supplementary English Glossary. Containing 12,000 Words and Meanings occurring in English Literature, not found in any other Dictionary. By Rev. T. L. O. Davies. Demy 8vo. 16*s.*

Folk-Etymology. A Dictionary of Words perverted in Form or Meaning by False Derivation or Mistaken Analogy. By Rev. A. S. Palmer. Demy 8vo. 21*s.*

Brief History of the English Language. By Prof. James Hadley, LL.D., Yale College. Fcap. 8vo. 1*s.*

The Elements of the English Language. By E. Adams, Ph.D. 20th Edition. Post 8vo. 4*s.* 6*d.*

Philological Essays. By T. H. Key, M.A., F.R.S. 8vo. 10*s.* 6*d.*

Language, its Origin and Development. By T. H. Key, M.A., F.R.S. 8vo. 14*s.*

Synonyms and Antonyms of the English Language. By Archdeacon Smith. 2nd Edition. Post 8vo. 5*s.*

Synonyms Discriminated. By Archdeacon Smith. Demy 8vo. 2nd Edition revised. 14*s.*

Bible English. Chapters on Words and Phrases in the Bible and Prayer Book. By Rev. T. L. O. Davies. 5*s.*

The Queen's English. A Manual of Idiom and Usage. By the late Dean Alford 6th Edition. Fcap. 8vo. 5*s.*

A History of English Rhythms. By Edwin Guest, M.A., D.C.L., LL.D. New Edition, by Professor W. W. Skeat. Demy 8vo. 18*s.*

Etymological Glossary of nearly 2500 English Words derived from the Greek. By the Rev. E. J. Boyce. Fcap. 8vo. 3*s.* 6*d.*

A Syriac Grammar. By G. Phillips, D.D. 3rd Edition, enlarged. 8vo. 7*s.* 6*d.*

DIVINITY, MORAL PHILOSOPHY, &c.

Novum Testamentum Græcum, Textus Stephanici, 1550. By F. H. Scrivener, A.M., LL.D., D.C.L. New Edition. 16mo. 4*s.* 6*d.* Also on Writing Paper, with Wide Margin. Half-bound. 12*s.*

By the same Author.

Codex Bezæ Cantabrigiensis. 4to. 26*s.*

A Plain Introduction to the Criticism of the New Testament. With Forty Facsimiles from Ancient Manuscripts. 3rd Edition. 8vo. 18*s.*

Six Lectures on the Text of the New Testament. For English Readers. Crown 8vo. 6*s.*

A Guide to the Textual Criticism of the New Testament.
By the Rev. Edward Miller, M.A. Crown 8vo. 4s.

The New Testament for English Readers. By the late H. Alford,
D.D. Vol. I. Part I. 3rd Edit. 12s. Vol. I. Part II. 2nd Edit. 10s. 6d.
Vol. II. Part I. 2nd Edit. 16s. Vol. II. Part II. 2nd Edit. 16s.

The Greek Testament. By the late H. Alford, D.D. Vol. I. 6th
Edit. 1l. 8s. Vol. II. 6th Edit. 1l. 4s. Vol. III. 5th Edit. 18s. Vol. IV.
Part I. 4th Edit. 18s. Vol. IV. Part II. 4th Edit. 14s. Vol. IV. 1l. 12s.

Companion to the Greek Testament. By A. C. Barrett, M.A.
5th Edition, revised. Fcap. 8vo. 5s.

The Book of Psalms. A New Translation, with Introductions, &c.
By the Very Rev. J. J. Stewart Perowne, D.D. 8vo. Vol. I. 5th Edition,
18s. Vol. II. 5th Edit. 16s.

—— Abridged for Schools. 5th Edition. Crown 8vo. 10s. 6d.

History of the Articles of Religion. By C. H. Hardwick. 3rd
Edition. Post 8vo. 5s.

History of the Creeds. By J. R. Lumby, D.D. 2nd Edition.
Crown 8vo. 7s. 6d.

Pearson on the Creed. Carefully printed from an early edition.
With Analysis and Index by E. Walford, M.A. Post 8vo. 5s.

Liturgies and Offices of the Church, for the use of English
Readers, in Illustration of the Book of Common Prayer. By the Rev.
Edward Burbidge, M.A. Crown 8vo. 9s.

An Historical and Explanatory Treatise on the Book of
Common Prayer By Rev. W. G. Humphry, B.D. 6th Edition, enlarged.
Small Post 8vo. 2s. 6d. ; Cheap Edition, 1s.

A Commentary on the Gospels, Epistles, and Acts of the
Apostles. By Rev. W. Denton, A.M. New Edition. 7 vols. 8vo. 18s.
each, except Vol. II. of the Acts, 14s. Sold separately.

Notes on the Catechism. By Rt. Rev. Bishop Barry. 7th Edit.
Fcap. 2s.

Catechetical Hints and Helps. By Rev. E. J. Boyce, M.A. 4th
Edition, revised. Fcap. 2s. 6d.

Examination Papers on Religious Instruction. By Rev. E. J.
Boyce. Sewed. 1s. 6d.

The Winton Church Catechist. Questions and Answers on the
Teaching of the Church Catechism. By the late Rev. J. S. B. Monsell,
LL.D. 4th Edition. Cloth, 3s. ; or in Four Parts, sewed.

The Church Teacher's Manual of Christian Instruction. By
Rev. M. F. Sadler. 34th Thousand. 2s. 6d.

Easy Lessons on the Church Catechism, for Sunday Schools.
By Rev. B. T. Barnes. 1s. 6d.

Butler's Analogy of Religion; with Introduction and Index by
Rev. Dr. Steere. New Edition. Fcap. 3s. 6d.

Kent's Commentary on International Law. By J. T. Abdy,
LL.D. New and Cheap Edition. Crown 8vo. 10s. 6d.

A Manual of the Roman Civil Law. By G. Leapingwell, LL.D.
8vo. 12s.

Essays on some Disputed Questions in Modern International
Law. By T. J. Lawrence, M.A., LL.M. 2nd Edition, revised and enlarged. Post 8vo. 6s.

FOREIGN CLASSICS.

A Series for use in Schools, with English Notes, grammatical and explanatory, and renderings of difficult idiomatic expressions.
Fcap. 8vo.

Schiller's Wallenstein. By Dr. A. Buchheim. 5th Edit. 5*s.*
Or the Lager and Piccolomini, 2*s.* 6*d.* Wallenstein's Tod, 2*s.* 6*d.*
—— **Maid of Orleans.** By Dr. W. Wagner. 1*s.* 6*d.*
—— **Maria Stuart.** By V. Kastner. 1*s.* 6*d.*

Goethe's Hermann and Dorothea. By E. Bell, M.A., and E. Wölfel. 1*s.* 6*d.*

German Ballads, from Uhland, Goethe, and Schiller. By C. L. Bielefeld. 3rd Edition. 1*s.* 6*d.*

Charles XII., par Voltaire. By L. Direy. 4th Edition. 1*s.* 6*d.*

Aventures de Télémaque, par Fénélon. By C. J. Delille. 4th Edition. 2*s.* 6*d.*

Select Fables of La Fontaine. By F. E. A. Gasc. 18th Edit. 1*s.* 6*d.*

Piociola, by X. B. Saintine. By Dr. Dubuc. 15th Thousand. 1*s.* 6*d.*

Lamartine's Le Tailleur de Pierres de Saint-Point. Edited, with Notes, by J. Boïelle, Senior French Master, Dulwich College. 2nd Edition. Fcap. 8vo. 1*s.* 6*d.* ——

Italian Primer. By Rev. A. C. Clapin, M.A. Fcap. 8vo. 1*s.*

FRENCH CLASS-BOOKS.

French Grammar for Public Schools. By Rev. A. C. Clapin, M.A. Fcap. 8vo. 11th Edition, revised. 2*s.* 6*d.*

French Primer. By Rev. A. C. Clapin, M.A. Fcap. 8vo. 6th Edit. 1*s.*

Primer of French Philology. By Rev. A. C. Clapin. Fcap. 8vo. 2nd Edit. 1*s.*

Le Nouveau Trésor; or, French Student's Companion. By M. E. S. 18th Edition. Fcap. 8vo. 1*s.* 6*d.*

French Examination Papers in Miscellaneous Grammar and Idioms. Compiled by A. M. M. Stedman, M.A. Crown 8vo. 2*s.* 6*d.*

Manual of French Prosody. By Arthur Gosset, M.A. Crown 8vo. 3*s.*

F. E. A. GASC'S FRENCH COURSE.

First French Book. Fcap. 8vo. 96th Thousand. 1*s.* 6*d.*

Second French Book. 42nd Thousand. Fcap. 8vo. 2*s.* 6*d.*

Key to First and Second French Books. 4th Edit. Fcp. 8vo. 3*s.* 6*d.*

French Fables for Beginners, in Prose, with Index. 15th Thousand. 12mo. 2*s.*

Select Fables of La Fontaine. New Edition. Fcap. 8vo. 3*s.*

Histoires Amusantes et Instructives. With Notes. 15th Thousand. Fcap. 8vo. 2*s.* 6*d.*

Practical Guide to Modern French Conversation. 15th Thousand. Fcap. 8vo. 2*s.* 6*d.*

French Poetry for the Young. With Notes. 5th Edition. Fcap. 8vo. 2*s.*

Materials for French Prose Composition; or, Selections from
the best English Prose Writers. 17th Thousand. Fcap. 8vo. 4s. 6d.
Key, 6s.

Prosateurs Contemporains. With Notes. 9th Edition, revised. 12mo. 5s.

Le Petit Compagnon; a French Talk-Book for Little Children.
11th Thousand. 16mo. 2s. 6d.

An Improved Modern Pocket Dictionary of the French and
English Languages. 36th Thousand, with Additions. 16mo. Cloth. 4s.
Also in 2 vols. in neat leatherette, 5s.

Modern French-English and English-French Dictionary. 3rd
and Cheaper Edition, revised. In 1 vol. 10s. 6d.

The A B C Tourists' French Interpreter of all Immediate
Wants. By F. E. A. Gasc. 1s.

GOMBERT'S FRENCH DRAMA.

Being a Selection of the best Tragedies and Comedies of Molière,
Racine, Corneille, and Voltaire. With Arguments and Notes by A.
Gombert. New Edition, revised by F. E. A. Gasc. Fcap. 8vo. 1s. each;
sewed, 6d.
                             CONTENTS.

MOLIERE:—Le Misanthrope. L'Avare. Le Bourgeois Gentilhomme. Le
Tartuffe. Le Malade Imaginaire. Les Femmes Savantes. Les Fourberies
de Scapin. Les Précieuses Ridicules. L'Ecole des Femmes. L'Ecole des
Maris. Le Médecin malgré Lui.

RACINE:—Phédre. Esther. Athalie. Iphigénie. Les Plaideurs. La
Thébaïde; ou, Les Frères Ennemis. Andromaque. Britannicus.

P. CORNEILLE:—Le Cid. Horace. Cinna. Polyeucte.

VOLTAIRE:—Zaïre.

GERMAN CLASS-BOOKS.

Materials for German Prose Composition. By Dr. Buchheim
10th Edition, thoroughly revised. Fcap. 4s. 6d. Key, Parts I. and II., 3s.
Parts III. and IV., 4s.

Advanced German Course. Comprising Materials for Translation, Grammar, and Conversation. By F. Lange, Ph.D., Professor
R. M. A. Woolwich. Crown 8vo. 1s. 6d.

GERMAN SCHOOL CLASSICS.

Meister Martin, der Küfner. Erzählung von E. T. A. Hoffman.
Edited by F. Lange, Ph.D., Professor, Royal Military Academy, Woolwich.
Fcap. 8vo. 1s. 6d.

Hans Lange. Schauspiel von Paul Heyse. Edited by A. A.
Macdonell, M.A., Ph.D., Taylorian Teacher, University, Oxford. *Authorised Edition*. Fcap. 8vo. 2s.

Auf Wache. Novelle von Berthold Auerbach. DER GEFRORENE
KUSS. Novelle von Otto Roquette. Edited by A. A. Macdonell, M.A.
Authorised Edition, fcap. 8vo. 2s.

Wortfolge, or Rules and Exercises on the Order of Words in
German Sentences. By Dr. F. Stock. 1s. 6d.

A German Grammar for Public Schools. By the Rev. A. C.
Clapin and F. Holl Müller. 3rd Edition. Fcap. 2s. 6d.

A German Primer, with Exercises. By Rev. A. C. Clapin. 1s.

Kotzebue's Der Gefangene. With Notes by Dr. W. Stromberg. 1s.

ENGLISH CLASS-BOOKS.

A Brief History of the English Language. By Prof. Jas. Hadley, LL.D., of Yale College. Fcap. 8vo. 1s.

The Elements of the English Language. By E. Adams, Ph.D. 20th Edition. Post 8vo. 4s. 6d.

The Rudiments of English Grammar and Analysis. By E. Adams, Ph.D. 15th Thousand. Fcap. 8vo. 2s.

A Concise System of Parsing. By L. E. Adams, B.A. Fcap. 8vo. 1s. 6d.

By C. P. MASON, Fellow of Univ. Coll. London.

First Notions of Grammar for Young Learners. Fcap. 8vo. 21st Thousand. Cloth. 8d.

First Steps in English Grammar for Junior Classes. Demy 18mo. 38th Thousand. 1s.

Outlines of English Grammar for the use of Junior Classes. 53rd Thousand. Crown 8vo. 2s.

English Grammar, including the Principles of Grammatical Analysis. 28th Edition. 110th Thousand. Crown 8vo. 3s. 6d.

A Shorter English Grammar, with copious Exercises. 21st Thousand. Crown 8vo. 3s. 6d.

English Grammar Practice, being the Exercises separately. 1s.

Code Standard Grammars. Parts I. and II. 2d. each. Parts III., IV., and V., 3d. each.

Practical Hints on Teaching. By Rev. J. Menet, M.A. 6th Edit. revised. Crown 8vo. cloth, 2s. 6d. ; paper, 2s.

How to Earn the Merit Grant. A Manual of School Management. By H. Major, B.A., B.Sc. 2nd Edit. revised. Part I. Infant School, 3s. Part II. 4s. Complete, 6s.

Test Lessons in Dictation. 3rd Edition. Paper cover, 1s. 6d.

Drawing Copies. By P. H. Delamotte. Oblong 8vo. 12s. Sold also in parts at 1s. each.

Poetry for the Schoolroom. New Edition. Fcap. 8vo. 1s. 6d.

The Botanist's Pocket-Book. With a copious Index. By W. R. Hayward. 4th Edit. revised. Crown 8vo. cloth limp. 4s. 6d.

Experimental Chemistry, founded on the Work of Dr. Stöckhardt. By C. W. Heaton. Post 8vo. 5s.

Picture School-Books. In Simple Language, with numerous Illustrations. Royal 16mo.

The Infant's Primer. 3d.—School Primer. 6d.—School Reader. By J. Tilleard. 1s.—Poetry Book for Schools. 1s.—The Life of Joseph. 1s.—The Scripture Parables. By the Rev. J. E. Clarke. 1s.—The Scripture Miracles. By the Rev. J. E. Clarke. 1s.—The New Testament History. By the Rev. J. G. Wood, M.A. 1s.—The Old Testament History. By the Rev. J. G. Wood, M.A. 1s.—The Story of Bunyan's Pilgrim's Progress. 1s.—The Life of Christopher Columbus. By Sarah Crompton. 1s.—The Life of Martin Luther. By Sarah Crompton. 1s.

BOOKS FOR YOUNG READERS.

A Series of Reading Books designed to facilitate the acquisition of the power of Reading by very young Children. In 9 vols. limp cloth, 6d. each.

The Old Boathouse. Bell and Fan ; or, A Cold Dip.
Tot and the Cat. A Bit of Cake. The Jay. The
   Black Hen's Nest. Tom and Ned. Mrs. Bee.
The Cat and the Hen. Sam and his Dog Red-leg.
   Bob and Tom Lee. A Wreck.
The New-born Lamb. The Rosewood Box. Poor
   Fan. Sheep Dog.

> *Suitable for Infants.*

The Story of Three Monkeys.
Story of a Cat. Told by Herself.
The Blind Boy. The Mute Girl. A New Tale of
   Babes in a Wood.
The Dey and the Knight. The New Bank Note.
   The Royal Visit. A King's Walk on a Winter's Day.
Queen Bee and Busy Bee.
Gull's Crag.
A First Book of Geography. By the Rev. C. A. Johns.
   Illustrated. Double size, 1s.

> *Suitable for Standards I. & II.*

BELL'S READING-BOOKS.

FOR SCHOOLS AND PAROCHIAL LIBRARIES.

The popularity of the 'Books for Young Readers' is a sufficient proof tha teachers and pupils alike approve of the use of interesting stories, in place of the dry combination of letters and syllables, of which elementary reading-books generally consist. The Publishers have therefore thought it advisable to extend the application of this principle to books adapted for more advanced readers.

Now Ready. Post 8vo. Strongly bound in cloth, 1s. each.

Grimm's German Tales. (Selected.)
Andersen's Danish Tales. Illustrated. (Selected.)
Great Englishmen. Short Lives for Young Children.
Great Englishwomen. Short Lives of.
Great Scotsmen. Short Lives of.
Masterman Ready. By Capt. Marryat. Illus. (Abgd.)

> *Suitable for Standards III. & IV.*

Friends in Fur and Feathers. By Gwynfryn.
Parables from Nature. (Selected.) By Mrs. Gatty.
Lamb's Tales from Shakespeare. (Selected.)
Edgeworth's Tales. (A Selection.)
Gulliver's Travels. (Abridged.)
Robinson Crusoe. Illustrated.
Arabian Nights. (A Selection Rewritten.)
Light of Truth. By Mrs. Gatty.

> *Standards IV. & V.*

The Vicar of Wakefield.
Settlers in Canada. By Capt. Marryat. (Abridged.)
Marie: Glimpses of Life in France. By A. R. Ellis.
Poetry for Boys. Selected by D. Munro.
Southey's Life of Nelson. (Abridged.)
Life of the Duke of Wellington, with Maps and Plans.
Sir Roger de Coverley and other Essays from the
   Spectator.
Tales of the Coast. By J. Runciman.
       Others in preparation.

> *Standards V. VI. & VII.*

London : Printed by STRANGEWAYS & SONS, Tower Street, St. Martin's Lane.

www.ingramcontent.com/pod-product-compliance
Lightning Source LLC
Chambersburg PA
CBHW021941190326
41519CB00009B/1089